LIZARDS
OF THE WORLD

LIZARDS OF THE WORLD

A GUIDE TO EVERY FAMILY

Mark O'Shea

PRINCETON UNIVERSITY PRESS
PRINCETON AND OXFORD

First published in the United States and Canada in 2021 by
Princeton University Press
41 William Street
Princeton, New Jersey 08540
press.princeton.edu

Copyright © 2021 Quarto Publishing plc

Conceived, designed, and produced by
Ivy Press
an imprint of The Quarto Group
The Old Brewery, 6 Blundell Street,
London N7 9BH, United Kingdom
T (0) 20 7700 6700
www.QuartoKnows.com

All rights reserved. No part of this book may be reproduced or transmitted in any form or by any means, electronic or mechanical, including photocopying, recording, or by any information storage-and-retrieval system, without written permission from the copyright holder.

Library of Congress Control Number: 2020948791
ISBN: 978-0-691-19869-9
Ebook ISBN: 978-0-691-21183-1

Publisher James Evans
Editorial Directors Tom Kitch, Isheeta Mustafi
Art Director James Lawrence
Managing Editor Jacqui Sayers
Commissioning Editor Kate Shanahan
Project Editors Caroline Earle, Joanna Bentley
Design Wayne Blades
Picture Research Kate Duncan, Polly Goodman
Illustrator John Woodcock

Cover photos: Front cover, clockwise from top left: Shutterstock/fivespots, /Spok83, /PetlinDmitry, /PetlinDmitry; Adobe Stock/Andrew Burgess; Shutterstock/Patrick K. Campbell, /Eric Isselee, /PetlinDmitry, /bluedog studio, /Will Thomass, /Pan Xunbin, /Eric Isselee. Spine: Shutterstock/bluedog studio. Back cover: Shutterstock/fivespots.

Printed in Singapore

10 9 8 7 6 5 4 3 2 1

6 Introduction

CONTENTS

86 The Lizard Infraorders

Sphenodontia & Dibamia

90 Gekkota

Scincomorpha

140 Lacertoidea

166 Iguania

206 Anguimorpha

232 Glossary
234 Resources
236 Index
240 Acknowledgments & Picture Credits

RIGHT | A brightly colored Madagascan Day Gecko, (*Phelsuma madagascarensis*), belying the idea that all geckos are nocturnal.

INTRODUCTION

Lizards are ectothermic ("cold-blooded") reptiles that are characterized by a body covering of overlapping scales. Most have four limbs with five digits on each foot, some are brightly colored and others are cryptically patterned, and many have excellent vision, a keen sense of smell, or acute hearing, while others have sacrificed all of this for life underground.

Lizards may be small ($2/3$ in/17 mm snout-to-vent length, or SVL) or large (up to 5 ft/1.54 m SVL). Most are nonvenomous, but some possess venom glands and a few can deliver a potentially dangerous, and certainly painful, venomous bite, while the claws, teeth, and tails of large species are also formidable weapons. Many small lizards can autotomize (discard) their tails in defense.

Lizards possess a transversely positioned cloaca (common genital-excretory opening). Most lay eggs, but many give birth to live young. Some use virgin birth, females reproducing without a male. Males of sexual species possess paired copulatory organs known as hemipenes. Lizards may be insectivorous, carnivorous, omnivorous, or herbivorous. They have adopted terrestrial, fossorial (burrowing), arboreal, and aquatic lifestyles, and may be diurnal, crepuscular, or nocturnal in activity.

Many lizards are extremely common. A total of 53 species have been recorded from a single sandridge site in the Great Victoria Desert of Australia, while individual lizard species may be found in almost plague proportions on small Pacific islands, presumably those with low

predator densities. The dry riverbeds in one part of southeastern New Guinea that seemingly lacks lizard-eating death adders (*Acanthophis*) can be so alive with Eastern Four-fingered Skinks (*Carlia eothen*) that the rustling of hundreds of them running over the leaf litter is audible some distance away. It is hardly surprising, given the frequent abundance of highly visible lizards, that researchers working on the much more secretive snakes have coined the term "lizard envy."

However, some lizards are rarely seen, due to their secretive lifestyles, and others may be rare, pushed to the brink of extinction by human activities, habitat loss, invasive predatory mammals or snakes, and climate change. Over a dozen species have gone extinct in the last 100 years, with dozens more in danger of extinction.

Today squamate reptiles (the lizards, worm-lizards, and snakes—but especially lizards) are amongst the most diverse and successful terrestrial vertebrates on our planet. They have taken advantage of almost every habitat available, from rainforest to desert, Arctic tundra to Pacific atolls and high mountain slopes, and even our homes. They are one of nature's great success stories, so it is no surprise that there are almost 7,000 species of lizards, over 300 species of worm-lizards, and over 3,800 species of snakes recognized, with new species being described on a regular basis (for example, 2020: 105 new lizards, one worm-lizard, and 38 snakes; 2019: 80 new lizards, one worm-lizard, 36 snakes). This book will focus on the lizards and worm-lizards, as well as their distinct cousin, the Tuatara.

EVOLUTION AND ORIGINS OF LIZARDS

Our planet was once ruled by the reptiles, the very word "Jurassic" conjuring up images of huge predatory *Tyrannosaurus rex* and intelligent, pack-hunting velociraptors. The Jurassic period lasted from 201 to 145 MYA (million years ago), but *Velociraptor mongoliensis* did not appear until the Late Cretaceous 75–71 MYA, in central Eurasia, and "Sue," the famous *T. rex* now in the Field Museum of Natural History, Chicago, did not cast her shadow over Laurentia (North America) until 68–66 MYA, just prior to the Cretaceous–Paleogene (K–Pg) extinction event. So the two species never met, and neither was actually around during the Jurassic period. But lizards were.

THE FOSSIL RECORD

The Tuatara (*Sphenodon punctatus*) and the squamate reptiles (snakes, lizards, and worm-lizards) are the surviving representatives of the Lepidosauria, the "scaled lizards." Various groups of extinct marine reptiles have also been cited as lepidosaurians, including the armored placodonts, long-necked plesiosaurs, and short-necked pliosaurs, together grouped as the Euryapsids, and another group of extinct marine reptiles, the mosasaurs, which were considered to be close to snakes.

The once globally distributed order Rhynchocephalia is represented in the fossil record from the Middle Triassic (240 MYA), but it went extinct in the northern supercontinent of Laurasia between the Late Jurassic and Early Cretaceous (160.5–100.5 MYA), possibly unable to compete with more adaptable squamates. The group hung on in South America until the Late Cretaceous (66 MYA), but only the Tuatara of New Zealand is extant (still alive today), with a fossil record dating back 16 million years.

The oldest known lizard fossil was discovered as recently as 1999, in the Italian Alps. It was described in 2003 as *Megachirella wachtleri*, after its "large hands" ("mega" = large, "chirella" = hands) and its collector, and dated at 240 MYA (Middle Triassic). However, it belongs to an extinct lineage with no living descendants. A reportedly Late Triassic iguanian fossil from India (*Tikiguania estesi*) has been dismissed as of Quaternary or Late Tertiary origin due to its close resemblance to extant agamids.

RIGHT | The fossilized remains of *Megachirella wachtleri*, discovered in 1999 in the Italian Alps, and described in 2003. Visible (bottom left) are the head, anterior body, and front limbs.

ABOVE | An artist's impression of the oldest known lizard, *Megachirella wachtleri*, from the Middle Triassic, possibly attempting to capture an early insect.

Other ancient fossils are known. An isolated skull dated to the Late Permian–Early Triassic (250 MYA) was collected in southern Africa and named *Paliguana whitei*, and another fossil, named *Kudnu mackinleyi*, was collected in Australia and also dated as Early Triassic. Both of these fossils belong to the extinct family Paliguanidae, and although their precise relationships to modern lizards cannot be determined, they were probably squamates and they suggest a Gondwanan origin for lizards. There is then a considerable gap before the first appearance of modern lizards in the fossil record. Most modern reptile lineages are thought to have evolved between the Early Jurassic (200–170 MYA) and Early Cretaceous (145–130 MYA), so lizards were certainly scuttling around the feet of Jurassic dinosaurs.

EARLY LIZARDS

Early lizards would have flourished in Jurassic and Cretaceous landscapes. Arthropods evolved 570 MYA, so there would have been an abundant supply of insects, crustaceans, and arachnids to feed on, while larger meat-eating species could have fed on smaller lizards or the rodent-like mammals that started to appear 200 MYA, or scavenged from carnivorous dinosaur kills. The climate was warm and wet, and vegetation flourished, providing plentiful food for herbivorous species, while verdant jungles and humid swamps represented an abundance of opportunities for diversification.

PLATE TECTONICS AND TIMELINES

When the earliest rhynchocephalians and squamates first appeared, around 250–240 MYA, they evolved at a time when today's continents were fused together as the supercontinent Pangaea. Global movements of species would have been much easier than they are today, when the landmasses are separated by vast expanses of ocean. This could account for the once global distribution of ancient groups like the rhynchocephalians.

Pangaea began to break apart during the Late Triassic to the Middle Jurassic (215–175 MYA) as the northern supercontinent, Laurasia, slowly separated from the southern supercontinent, Gondwana. Modern squamates were beginning to appear, but there was still time for some of the older groups to radiate across both landmasses. Those lizard groups that appeared before the breakup of Pangaea could be expected to occur in the fossil records of both Laurasia and Gondwana.

During the Late Triassic–Early Jurassic (200 MYA) Laurasia was also breaking apart, as Laurentia (North America and Greenland) began to separate from Eurasia, leading to the creation of the North Atlantic Ocean. Although these two landmasses were moving apart, North Atlantic land bridges are thought to have existed twice, during the Late Jurassic (c.154 MYA) and Early Cretaceous (c.131 MYA), possibly accounting for the transatlantic distributions of some lizard taxa.

During the Jurassic, Cretaceous, and Early Paleogene Periods (180–45 MYA), Gondwana split into South America, Africa, Madagascar, the Seychelles, India, Australia, and Antarctica, which then moved apart through a process known as

THE SEPARATION OF THE SUPERCONTINENTS OVER THE PAST 225 MILLION YEARS

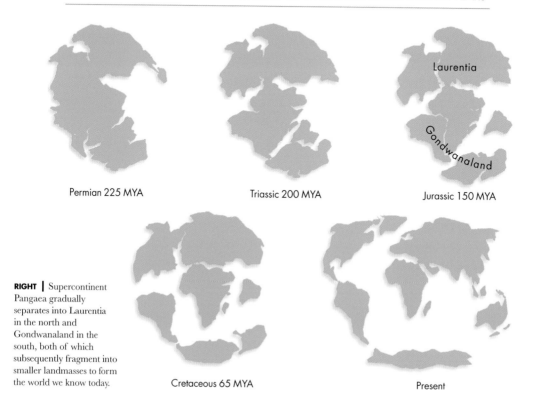

Permian 225 MYA

Triassic 200 MYA

Jurassic 150 MYA

Cretaceous 65 MYA

Present

RIGHT | Supercontinent Pangaea gradually separates into Laurentia in the north and Gondwanaland in the south, both of which subsequently fragment into smaller landmasses to form the world we know today.

BELOW | The Megalania (*Varanus priscus*), the largest known lizard in Pleistocene Australia, went extinct around 48,000 MYA.

continental drift. India moved north and collided with Eurasia, the collision forming the Himalayas, while Arabia broke from Africa to collide with western Eurasia. Madagascar remained isolated, accounting for its high degree of endemicity, something also visible in the microcontinent of the Seychelles despite much of it sinking beneath the Indian Ocean to leave only mountaintops as islands. Much more recently, in the Neogene (3–2.8 MYA), the Isthmus of Panama formed to link North and South America and permit a mixing of their faunas. All of these plate movements, including the rifting, the collisions, and the isolation of islands, played a part in the modern-day distribution of lizards.

THE LARGEST LIZARD TO WALK THE EARTH

Today the largest lizard is the Komodo Dragon (*Varanus komodoensis*) from Indonesia (see page 230). A large male can achieve a snout-to-vent length (SVL) of 5 ft (1.5 m), a total length (TTL) of 10 ft (3 m), and a weight of 250 lb (113 kg), which is pretty impressive, but the Komodo Dragon is not the largest lizard to have stalked the Earth (remember, dinosaurs were not big lizards). A much larger lizard survived in Australia during the Pleistocene ice ages when sea levels were much lower and land bridges existed between Australia and New Guinea, and much of Indonesia was connected to the Southeast Asian mainland. That lizard was previously called *Megalania prisca*, but is now known as *Varanus priscus* due to its close relationship to other large Australian monitor lizards such as the Perentie (*V. giganteus*) and the Lace Monitor (*V. varius*).

The size of the Megalania has been the source of some disagreement, with TTLs of 18–26 ft (5.5–7.9 m) proposed, and weights ranging from 705 lb (320 kg) if built like *V. varius* to 4,280 lb (1,940 kg) if built like *V. komodoensis*. When Megalania went extinct is not known, but estimates place it around 48,000 years ago. Humans arrived in Australia between 65,000 and 40,000 years ago and were potentially responsible for the extinction of the Megalania, just as they were responsible for the extinction of other megafauna in Australia and elsewhere.

RIGHT | A family tree for the lizards, worm-lizards, and Tuatara, illustrating one proposed phylogeny for the infraorders Dibamia, Gekkota, Scincomorpha, Lacertoidea, Anguimorpha, and Iguania, and indicating the position of the Toxicofera clade (see below). The Rhynchocephalia contains one living family, Sphenodontidae, and one living species, the Tuatara (*Sphenodon punctatus*, page 88), the sister-clade to the entire Squamata. The Squamata (excluding snakes) contains 43 families, 13 of which contain between two and nine subfamilies. This is a simplified family tree, and the lengths of the arms are not intended to indicate the timelines since divergence between the various groups. Extinct taxa and lineages have been omitted. This tree is based on several published phylogenetic trees, including Zheng & Wiens (2016) *Molecular Phylogenetics and Evolution* 94:542.

EVOLUTIONARY TREES AND REOCCURRING TRENDS

There are a number of different evolutionary histories proposed for modern squamate reptiles, especially with molecular techniques now questioning earlier theories. One has the Iguania at the base of the squamate phylogenetic tree, as the sister clade to the Scleroglossa, which contains all other squamates. A rival phylogeny, and the one adopted here (see diagram opposite), places the Dibamia at the base of the squamate tree, with the Gekkota the next diverging branch, although other workers consider Gekkota more basal, and some place the Dibamia as the sister taxon to the Amphisbaenia, in the Lacertoidea. In 2005, Iguania was aligned with the Anguimorpha and the Serpentes (snakes) in Toxicofera, a clade (taxonomic group) that was defined as containing all squamates possessing "toxin-secreting oral glands" and based on the argument that venom evolved only once in the Squamata, albeit with some groups, like pythons, subsequently becoming nonvenomous again. However, the single venom evolution concept and the Toxicofera clade have not found universal acceptance.

Different phylogenetic arrangements have their supporters and detractors, but none of these arguments about ancestry, relationships, and evolution undermine the central fact that squamate reptiles are extremely diverse and very successful, and that the "Age of the Reptiles" did not end with the disappearance of the dinosaurs.

Throughout the evolutionary history of squamates there have been reoccurring trends that evolved independently multiple times, in unrelated lineages. The three most striking are the evolution of limblessness, usually in association with body elongation and a fossorial lifestyle (see page 20), the evolution of viviparity, or live-bearing, often as a response to life in the cooler climates that prevail at higher latitudes or altitudes (see page 55), and the evolution of herbivory (see page 58).

PHYLOGENY OF LIZARDS, WORM-LIZARDS, AND THE TUATARA

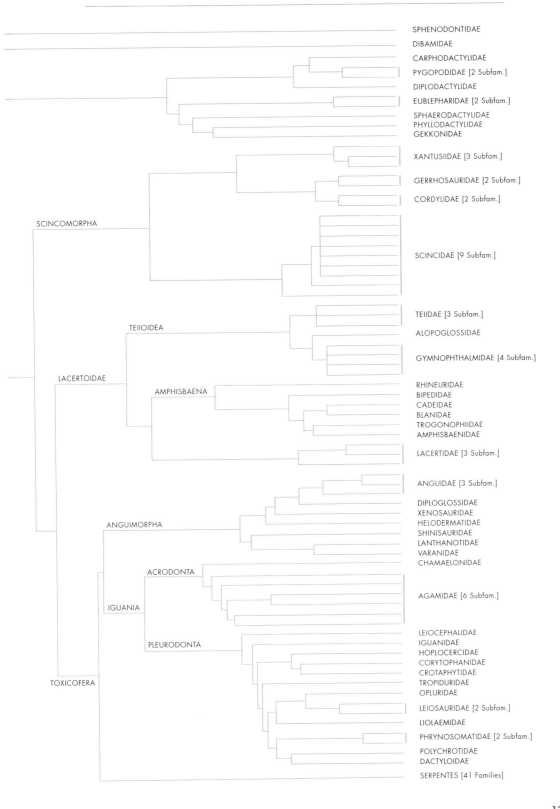

TAXONOMY OF LIZARDS

RIGHT & BELOW | The Synapsid skull of a proto-mammal (right) exhibits a single fenestra (opening) in the skull behind the eye on either side of the head. The Diapsid skull of a Nile crocodile (*Crocodylus niloticus*, below) exhibits two fenestrae in the skull behind the eye, one lateral, one dorsal, on either side of the head.

Humans have always sought to categorize the other animals and plants with which they share the planet. Our ancestors likely categorized large animals into three groups: those that were good to eat, those that were not good to eat, and those that would eat us. Even people living in remote tropical locations, isolated from Western science and education, developed their own names and system of categorization for the mammals, birds, snakes, lizards, and frogs they encountered on a daily basis. It is possible that being able to categorize other animals gave humans some sort of imagined power over nature, or it may be that we just have tidy minds and need to know "what goes with what." Either way, all life on Earth, both extant and extinct, has been the subject of taxonomy, a classification system using hierarchical categories that dates back to the Swedish naturalist Carl Linnaeus, and the tenth edition of his magnum opus *Systema Naturae*, published in 1758. With certain modifications and new methodologies, we still use this system today.

A NOTE ON TAXONOMY AND SCIENTIFIC NAMES

Living organisms are classified using a hierarchical system of categories known as clades. A clade is any natural monophyletic group of organisms, meaning it contains a common ancestor and all its descendants. For lizards these clades are ranked, from highest to lowest, as: Domain: Eukarya; Kingdom: Animalia; Phylum: Chordata; Superclass: Tetrapoda; Class: Reptilia; Order: Squamata. Not all clades have a rank (for example, Amniota). Within the Squamata, lizards are grouped into infraorders, families (suffix -idae), and subfamilies (suffix -inae). Within the families and subfamilies are the genera, which contain the species. A species name is a binomial, comprising two words, and is written in italics with only the

THE AMNIOTIC EGG

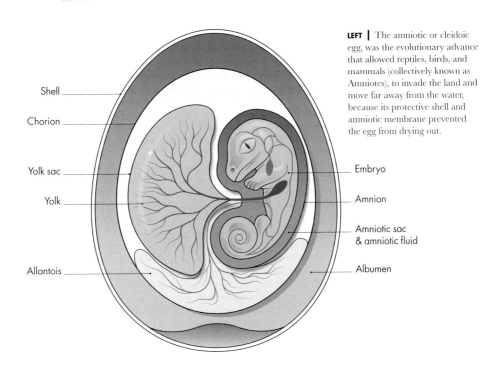

LEFT | The amniotic or cleidoic egg, was the evolutionary advance that allowed reptiles, birds, and mammals (collectively known as Amniotes), to invade the land and move far away from the water, because its protective shell and amniotic membrane prevented the egg from drying out.

first (generic) part receiving a capitalized initial letter (such as *Zootoca vivipara* for the Viviparous Lizard). A trinomial name (with three parts) indicates a subspecies. The binomial name may be accompanied by the name of the person who first described the species, and the date of publication of the description. If the author's name and date are contained in parentheses, this indicates that the species' name has changed since it was first described, usually because it has been moved to a different genus (for example, *Zootoca vivipara* (Lichtenstein, 1823) was originally described by Lichtenstein as *Lacerta vivipara*).

DEFINING LIZARDS

SUPERCLASS TETRAPODA: Modern amphibians, reptiles, birds, and mammals are classified as pentadactyl tetrapods (five-digited, four-limbed vertebrates), a clade that includes all vertebrates except fish. Any subsequent reduction or loss of digits or limbs, due to adoption of a fossorial or aquatic lifestyle (for example, in caecilians, snakes, slow worms, and marine mammals), does not disqualify an organism from belonging to this clade because its ancestors conformed to the pentadactyl tetrapod "blueprint."

UNRANKED CLADE AMNIOTA: Reptiles, birds, and mammals are separated from the amphibians because their ancestors evolved a waterproof skin, and the amniotic (or cleidoic) egg, which contains an amniotic membrane to prevent its desiccation, enabling it to be laid on land. Unlike amphibians, amniotes were not forced to return to freshwater to reproduce. First to branch away from the amniotic phylogenetic tree were the Synapsida, whose skulls possessed a single opening, a fenestra, posterior to the eye—this clade would evolve into the modern mammals.

RELATIONSHIPS OF THE REPTILIA

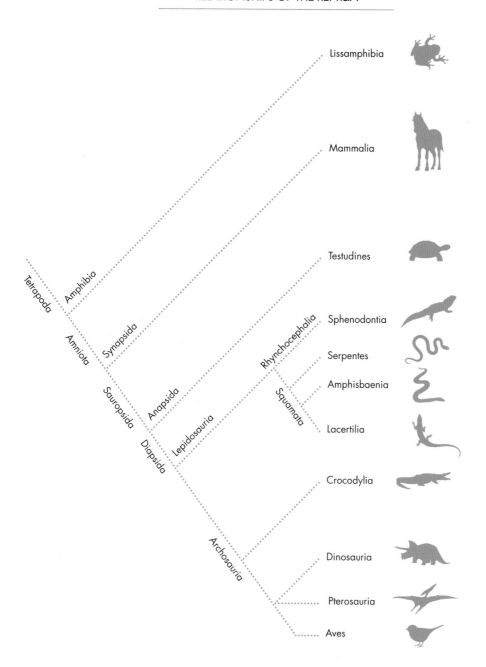

ABOVE | Lizards and worm-lizards are members of the Tetrapoda (four limbed vertebrates); the Amniota (vertebrates that produce an amniotic egg which do not need to return to water to reproduce); the Sauropsida (all reptiles and birds); the Diapsida (reptiles with two openings known as fenestrae on the rear of their heads); the superorder Lepidosauria (scaled reptiles); and the order Squamata (lizards, worm-lizards, and snakes).

UNRANKED CLADE SAUROPSIDA: The sister clade to the Synapsida, the Sauropsida contains all extant and extinct reptiles and birds. Traditionally the next to diverge were thought to be the Anapsida, ancestral turtles whose skulls lacked any lateral openings behind the eyes; they would evolve into the modern Testudines. However, the precise time of anapsid divergence is open to argument and it is thought they may be more closely related to crocodilians and may have diverged much later. All other reptiles and birds are diapsids; they possess two fenestrae in the lateral-posterior skull, one above the other.

CLASS REPTILIA: This is a rather artificial clade and difficult to define, since reptiles are traditionally identified more by what they are not than by what they are. What an average person defines as a reptile does not have slimy, semipermeable skin (amphibians), feathers (birds), or fur (mammals); their bodies may be covered in leathery skin (crocodiles), bony shells (turtles), or scales (snakes, lizards, worm-lizards, and the Tuatara). Unlike the endothermic ("warm-blooded") mammals and birds, all modern "reptiles" are ectothermic, gaining their body heat from the environment. The Reptilia, as traditionally recognized, is paraphyletic; that is, it does not contain a common ancestor and *all* its living descendants, because the birds are separated from it as the Aves. Since taxonomy prefers monophyletic groups that contain an ancestor and all its descendants, either the birds should be included in the Reptilia to maintain its monophyly, or the living Reptilia should be split into four separate clades: Testudines (turtles), Crocodylia, Aves, and Lepidosauria. The former seems to be the more popular, with authors referring to "non-avian reptiles" when they mean modern reptiles but not birds.

UNRANKED CLADE DIAPSIDA: The Diapsida is the Sauropsida (or the Reptilia) without the Anapsida (turtles), although recent research suggests that turtles may be diapsids that reverted to the anapsid condition. Diapsida comprises two ancestral lineages, the Archosauria ("ancient lizards"), which contains the dinosaurs, pterosaurs, crocodilians, and birds, and the Lepidosauria ("scaled lizards"). Lepidosaurs also exhibit a transverse cloaca, in contrast to crocodilians, in which the cloaca is longitudinally positioned.

SUPERORDER LEPIDOSAURIA: The Lepidosauria contains the Rhynchocephalia (beak-headed reptiles) and the Squamata. The Rhynchocephalia was once distributed globally, with both terrestrial and marine species, but today the order is represented by a single extant species, the Tuatara (*Sphenodon punctatus*) of New Zealand.

ORDER SQUAMATA: The order Squamata is the largest and most successful clade of living reptiles, with over 11,000 species. It has been traditionally split into three suborders: Lacertilia, or Sauria (lizards), Amphisbaenia (worm-lizards), and Serpentes, or Ophidia (snakes). However, the Lacertilia is paraphyletic because both worm-lizards and snakes evolved from within the lizards, just as birds evolved from within the reptiles; so the terms Lacertilia or Sauria have therefore fallen out of favor.

INTERNAL ANATOMY

The basic lizard skeleton comprises a skull, with a hinged mandible (lower jaw) that is fused at the chin (unlike in snakes, which can articulate their lower jaws); a vertebral column; a pectoral girdle and forelimbs, each with five digits; a rib cage that lacks a sternum (breastbone); a pelvic girdle and hind limbs, each with five digits; and, posteriorly, a tail. Across the combined total of more than 7,000 lizard and worm-lizard species there are considerable deviations from this basic lizard blueprint.

As an air-breathing vertebrate a lizard also possesses a pair of lungs, a three-chambered heart, two kidneys, a liver, pancreas, and gall bladder, circulatory, lymphatic, and nervous systems, and the reproductive organs requisite for its gender. The same soft tissue internal organs are required by all tetrapods, although their arrangement in a long, slender animal like a snake or legless lizard is different to the way they are arranged in the body of a typical lizard.

LIZARD SKELETONS

Life on land requires that a lizard's musculoskeletal system be robust, and strong enough to raise its body off the ground, support its weight by counteracting gravity, and move its body forward, often at speed when pursuing prey or escaping from a predator. Lizards lack a sternum, which means they can expand their ribs outward to flatten the body dorsoventrally when basking, as a means of territorial display in the colorful butterfly lizards (*Leiolepis*, page 174), or to form a patagium in gliding species (*Draco*, page 172).

COMPARISON OF MODERN REPTILE AND MAMMAL LIMBS

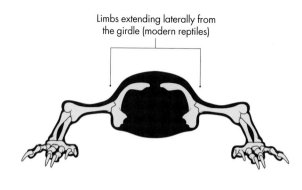

Limbs extending laterally from the girdle (modern reptiles)

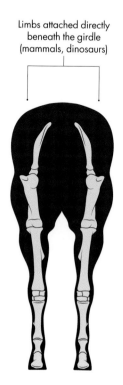

Limbs attached directly beneath the girdle (mammals, dinosaurs)

THE LIZARD SKELETON

The way the limbs are attached to the pectoral and pelvic girdles in lizards differs from how they are attached in mammals. In dinosaurs and mammals the limbs are erect, almost directly below the body, so when they run it is essentially in a straight line. However, in modern reptiles the limbs are sprawling, extending outward and down. In this position a running lizard throws its body into a series of S-shaped curves, the head and tail swinging from side to side in response to the curvature of the spinal column. Some lizards run bipedally (on two legs), which results in motion similar to a human running, albeit with a more gangling gait. Running and other forms of movement, and how lizards have adapted, are discussed under Locomotion (see page 40).

Lizard tails also demonstrate tremendous variation. Many lizards possess tails that are exceedingly long and slender, much longer than the lizard's SVL. The Papuan Monitor (*Varanus salvadorii*) has a tail approximately twice its SVL, while that of the Long-tailed Grass Lizard (*Takydromus sexlineatus*) is five times its SVL. Other lizards possess extremely short tails; for example, Leach's Giant Gecko (*Rhacodactylus leachianus*), which has a tail 18–24 percent of its SVL, or the Centralian Knob-tailed Gecko (*Nephrurus amyae*), whose tail is 15–25 percent of its SVL. Lizard tails may also be wider than the body, and covered in spines, for example in the Broad-tailed Gecko (*Phyllurus platurus*). Tails are used for balance, as climbing aids, for display, defense (caudal autotomy, page 73), as a weapon, or for blocking burrows, and different uses require different designs.

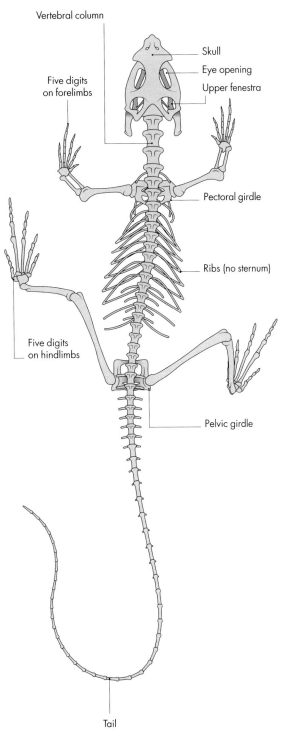

THE TREND TOWARD LEGLESSNESS

Four legs with five toes on each foot is the basic pentadactyl tetrapod blueprint, with the limbs long enough to overlap if they are pressed against the body, and the fourth digit on each foot the longest. A great many lizards comply with this pattern.

However, not all lizards are pentadactyl, or even four-legged. A trend toward leglessness can be seen across several unrelated lineages. This trend toward limb reduction first manifests as a loss of phalanges (the individual bones of the fingers or toes), resulting in shortened digits. There may then be subsequent loss of entire digits, reduction in limb length, and even complete loss of limbs, usually combined with body elongation, which ultimately leads to complete limblessness, as demonstrated by the snakes (except boas and pythons), the slow worms (*Anguis*), blind-lizards (*Dibamus*), many skinks, and most worm-lizards.

Limb reduction has evolved independently at least 54 times in the Squamata, and is present in virtually every squamate lineage, albeit subtly in some agamas and chameleons that have only lost a single phalange from the fourth finger or toe. It is much more apparent in those lineages where whole digits or limbs are absent. Within the Gekkota this includes the entire Pygopodidae, 46 species of elongate Australo-Papuan geckos which completely lack forelimbs, but possess vestigial hind limbs as scaly flaps that can serve little purpose for locomotion.

Scincomorphan limb reduction is represented in the normally four-legged girdled lizards (Cordylidae) by the grass lizards (*Chamaesaura*), with two species exhibiting single-digit fore- and hind limbs, and two species possessing only hind limbs. In the related plated lizards (Gerrhosauridae), the seps (*Tetradactylus*) also demonstrate body and tail elongation, limb reduction, and digit loss, the most extreme example being Breyer's Long-tailed Seps (*T. breyeri*), which has two-digit forelimbs and single-digit hind limbs. However, the Scincidae is the most diverse family exhibiting limb reduction; it has occurred at least 25 times, across every

BELOW | An Emerald Tree Skink (*Lamprolepis smaragdina*) showing the basic lizard blueprint, with four well-developed limbs, each with five digits, with the fourth toe and fourth finger usually the longest. The distance between the limbs is shorter than the combined total length of the limbs, which would therefore overlap if adpressed against the body.

ABOVE | A Large-scaled Grass Lizard (*Chamaesaura macrolepis*) with an elongate body, complete loss of forelimbs, hind limb reduced to a tiny flap (inset), and a tail that is three to four times the SVL. This lizard is perfectly adapted for "swimming" through long grass, with an occasional propulsive flick from the limb-flap.

subfamily. The lance skinks (Acontinae) are an entirely legless subfamily, while the burrowing skinks (Scincinae) and Australo-Asian skinks (Sphenomorphinae) contain species exhibiting every possible variation from pentadactyl tetrapod to complete limbnessness, with the Australian grassland and desert sliders (*Lerista*) even exhibiting every combination across the 99 species in the genus.

Among the Anguimorpha demonstrating limb reduction are the slow worms, Asian glass lizards (*Dopasia*), and American glass lizards (*Ophisaurus*), which are completely limbless, and the Scheltopusik (*Pseudopus apodus*) and Moroccan Glass Lizard (*Hyalosaurus koellikeri*), which possess scaly hind limb flaps. The Californian legless lizards (*Anniella*) and the South American legless lizards (*Ophiodes*) closely resemble European slow worms, although the South American legless lizards possess vestigial hind limbs.

In the Lacertilia, the microteiids (Gymnophthalmidae) demonstrate a wide variety of limb reduction patterns, from pentadactyl tetrapod to the limbless sand teiids (*Calyptommatus*). The microteiid genus exhibiting the greatest range is *Bachia*, from the Trinidad Bachia (*B. trinitatis*), with five fingers and four toes, to the Madeira River Bachia (*B. scaea*), with no hind limbs and the forelimbs reduced to tiny spikes.

Many lizards have lost their forelimbs but retained their hind limbs, but the reverse situation is extremely rare. The most obvious examples are the Mexican ajolotes (*Bipes*), three species with long, annular-scaled bodies, well-developed five-, four-, or three-fingered forelimbs, but no hind limbs. They are the only members of the Amphisbaenia to possess even vestigial limbs. Among the lizards (Lacertilia), only the Madagascan pink skinks (*Voeltzkowia*), the Thai Jarujin Skink (*Jarujinia bipedalis*), and six *Bachia* species have forelimbs but no hind limbs.

Extreme limb reduction and body elongation are often associated with a fossorial lifestyle, since limbs could be an encumbrance for a burrower, and other characteristics often bundled with subterranean life include loss of the external ear openings, no eyelids but instead a transparent, snake-like "spectacle" over the eye, and a shortened tail. However, those limbless lizards inhabiting grassland habitats, such as the grass lizards (*Chamaesaura*), often possess tails three to four times the length of the body.

LIZARD SKULLS

Squamate reptiles possess a diapsid skull, with two fenestrae on either side of the skull, posterior to the orbits (eye openings). The primitive condition, in ancestral squamates and the Tuatara, includes a lower temporal arch beneath the lowest of the

THE TEGU SKULL

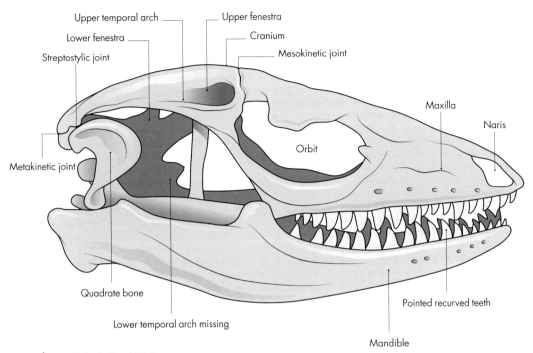

LEFT | Burton's Snake-lizard (*Lialis burtonis*) preying on a Bicarinate Four-fingered Skink (*Carlia bicarinata*) in southern Papua New Guinea, showing the cranial kinesis required to crush and manipulate such prey.

fenestrae. Modern squamates have lost this bone and the quadrate bones are streptostylic, which means they are able to rotate and articulate on the skull to move the mandible forward or backward.

The squamate cranium is capable of a high degree of kinesis (motion). It has flexible joints, making it possible for the bones to move in relationship to one another, changing the shape of the skull. Snakes have the greatest cranial movement, but many lizards are also able to flex their skull bones. The degree of movement depends on the lizard's diet and feeding technique; for example, it may increase the gape of the open mouth, or angle the snout downward as the mouth closes, exerting additional pressure and gripping the prey.

Chameleons have dispensed with this motion and possess a rigid cranium, although they do retain the streptostylic functionality of the quadrate bones. An extreme example of cranial movement can be seen in Burton's Snake-lizard (*Lialis burtonis*). This pygopodid lizard has an elongate snout which is capable of considerable movement that enables it to capture, manipulate, and swallow skinks that are relatively large in comparison to its own body size. Its feeding adaptations are convergent with those of the slender-snouted vinesnakes (*Oxybelis* and *Ahaetulla*), which also prey on lizards, and the teeth of *Lialis* are long and recurved like vinesnake teeth, not like those of typical invertebrate-eating gekkotans.

BELOW | The skull of a Guianan Caiman Lizard (*Dracaena guianensis*) showing a robust quadrate bone and molariform (molar-like) teeth for crushing snails.

LIZARD DENTITION

Squamate teeth may be pleurodont, arranged along the inner edges of the jawbones, or acrodont, arranged along the crest of the jaw ridge. Pleurodont is the ancestral condition, although rhynchocephalians exhibit the acrodont arrangement. Snakes have a modified pleurodont arrangement, the teeth being located in shallow sockets.

Snakes and lizards deal with food in entirely different ways. Snakes are obligate carnivores that overpower prey using constriction or venom, engulf it with their jaws, and swallow it whole. They are capable of ingesting large prey items because the lower jaw is not fused at the chin, and this separation also means they can alternately use the two halves of the lower jaw to draw prey down the throat, but at a sacrifice, as they cannot exert much bite force. Snakes' teeth are recurved and, apart from the modified fangs, relatively homogeneous, their purpose being to maintain contact with the prey and direct it down the throat.

Lizards may be carnivorous, herbivorous, or omnivorous and their dentition usually reflects their primary diet. Their mandibles are fused at the chin and hinged on the quadrate bones by powerful external jaw muscles. They are capable of exerting considerable bite force and delivering a crushing or shearing bite, reducing their meal to bite-sized chunks.

As well as being pleurodont (iguanids, geckos, and teiids) or acrodont (agamas and chameleons), lizard teeth may be recurved, blunt, bicuspid or tricuspid (with two or three points), or possess blade-like or serrated edges. Herbivorous lizards, such as iguanas, possess teeth that are angled and overlap slightly, enabling them to shear through vegetation. Durophagous lizards—those that feed on foods with hard exteriors, such as seeds, beetles, snails, crabs, or eggs, which must be crushed prior to ingestion—are equipped with blunt, crushing teeth, heavy-set mandibles, and powerful jaw muscles, the extreme example of this being a snail-specialist, the Guianan Caiman Lizard (*Dracaena guianensis*).

Worm-lizards (Amphisbaenia) are unique in that they possess a single, centrally arranged tooth in the front of the upper jaw that interlocks with a pair of teeth in the lower jaw to achieve a cutting, forceps-like grip on their prey of earthworms or insect larvae. The lateral teeth of both jaws also interlock when the mouth is closed.

LIZARD DENTITION

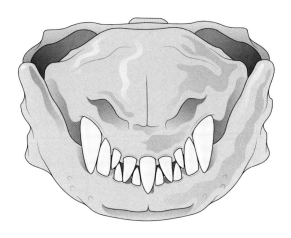

Amphisbaenid skull illustrating the unique centrally arranged tooth in the upper jaw

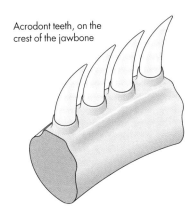

Acrodont teeth, on the crest of the jawbone

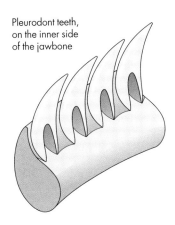

Pleurodont teeth, on the inner side of the jawbone

THORACIC ORGANS

As a terrestrial vertebrate, a lizard possesses virtually the same internal organs as a mammal or a bird, albeit based on an ectothermic (cold-blooded) metabolism. For those lizards that possess a typical tetrapod body shape, the organs will be symmetrically arranged, while an asymmetrical layout might be expected for lizards and worm-lizards with narrow, elongate, serpentine body shapes.

The most significant difference between the internal organs of lizards and those of mammals and birds is the three-chambered squamate heart, which comprises two atria and a single ventricle. The only extant non-avian reptiles with a four-chambered heart (with two ventricles) are the crocodilians, with turtles possessing a similar heart to the squamates. In lizards, the single ventricle contains three subchambers, the first two of which are separated by a muscular ridge, while the third is connected to one of the others via the intraventricular canal. The heart is therefore effectively five-chambered, but functions at any one time as four-chambered.

Lizards possess paired, symmetrical lungs that are connected to the outside world via the paired bronchi and the trachea, glottis, and mouth and nostrils. Squamates lack a secondary palate, which in mammals separates the nasal and oral cavities, the result being that the nasal cavity opens into the roof of the mouth. In mammals, the glottis is an opening in the larynx between the vocal chords. Normally open, it is closed by a flap called the epiglottis during the swallowing process to prevent ingress of food or liquid into the lungs. In squamates, the glottis appears as a visible slit-like opening posterior to the tongue and, in contrast to the arrangement in mammals, it is usually closed, only opening when the reptile takes a breath. As squamates lack a mammalian diaphragm, the

THE SQUAMATE HEART

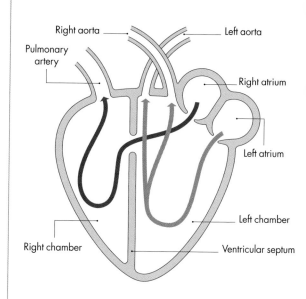

ABOVE | The squamate heart has two atria and a ventricle comprising three chambers, permitting right-to-left and left-to-right cardiac shunts and mixing of oxygenated and deoxygenated blood.

intercostal muscles that facilitate breathing are the same muscles used during locomotion, and they may be required to hold their breath during periods of intense activity.

Many lizards have valvular nostrils that can open or close to control entry of air into the buccal cavity. Respiration in lizards has three phases: expiration, inspiration, and relaxation. During expiration the lizard's nostrils and glottis are open as intercostal muscles compress the rib cage, raising the pressure in the lungs and forcing air out through the nostrils. During the next phase, inspiration, intercostal muscular activity expands the rib cage, resulting in negative pressure in the lungs, which draws in air through the nostrils and glottis, both of which then close. There is then a period of relaxation before the cycle of expiration/inspiration begins again. This process is called Negative Pressure Ventilation.

INTERNAL ANATOMY OF A (FEMALE) LIZARD

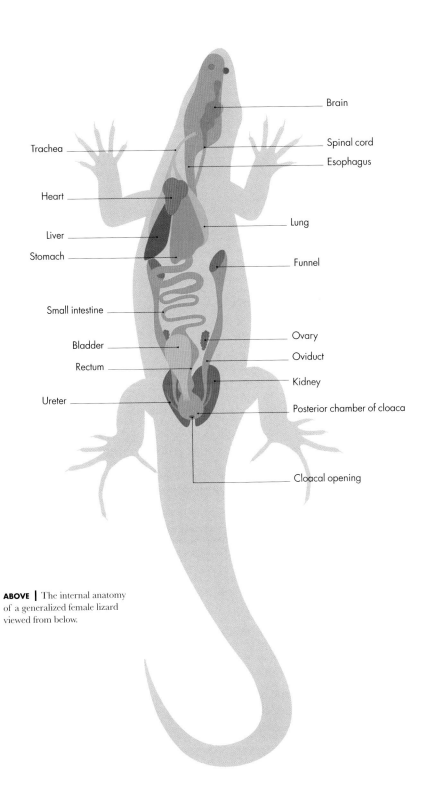

ABOVE | The internal anatomy of a generalized female lizard viewed from below.

LIZARD SKIN

The external integument (skin) of reptiles is one of the evolutionary adaptations that gave them their advantage over amphibians, which possess semipermeable skins that make them vulnerable to desiccation on land. Reptile skin also provides protection against injury, through the presence of scales, scutes, and osteoderms, and has specialized glands for territorial scent marking or leaving pheromonal trails. The scales of some lizards have been highly modified for purposes ranging from camouflage to gathering rainwater.

SKIN AND SCALES

Two types of keratin proteins are found in the skin of vertebrates: alpha-keratin is soft and elastic and found in all vertebrates, but beta-keratin is hard and rigid, and only present in the skin of reptiles and birds. It is the beta-keratin layer, overlaying the alpha-keratin layer and a meso-layer of mucus, that prevents water loss through reptilian skin. The outer surface of the beta-keratin layer becomes the "Oberhaütchen," a thin cuticle containing species-specific micro-ornamentations and skin sense organs. Squamate scales are produced through keratinization (thickening) and folding of this outer epidermis, with the outer cell layers dying off. Large, regular scales, especially those on the head, are often referred to as scutes.

While crocodilians and turtles grow continually throughout their active lives, snakes and lizards get larger through a series of alternating resting and growth periods. The latter is usually initiated by feeding or pregnancy, but also occasionally as a response to injury. Since the outer epidermal layers are dead, any body size increase will require them to be shed, a process known as ecdysis, which is initiated by the development of new Oberhaütchen, beta-keratin, meso, and alpha-keratin layers beneath the older layers. While snakes usually slough their entire skins in one go, lizards and the Tuatara shed their skins in patches. Body coloration and patterning is due to the presence of pigmented cells deeper in the dermis that are visible through the transparent epidermis. Therefore, sloughed skins usually only illustrate the faintest hint of the reptile's patterning.

SQUAMATE SKIN

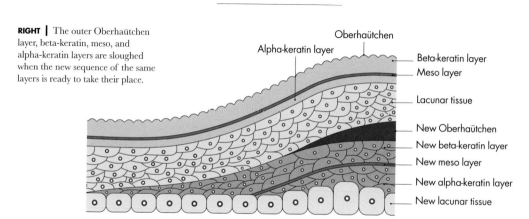

RIGHT | The outer Oberhaütchen layer, beta-keratin, meso, and alpha-keratin layers are sloughed when the new sequence of the same layers is ready to take their place.

The dorsal and lateral body scales of some squamates (skinks, girdled lizards, and alligator lizards) overlay strengthened structures in the skin known as osteoderms, literally "bone-skin." These osteoderms act as additional body armor for the lizard and, like medieval plated armor, they are only loosely attached to each other for flexibility during locomotion. Gila Monsters and beaded lizards (*Heloderma*) even possess osteoderms on their heads that are fused with the bones of the skull.

Many lizards, such as iguanids, exhibit a series of large glands known as femoral pores, each in the center of an enlarged scale on the posterior undersides of the hind limbs, and in some geckos and agamas these curved rows of pores are united by a series of pre-cloacal pores, anterior to the cloaca. Usually larger in males, they are used for scent marking. Those of the desert iguanas (*Dipsosaurus*) produce a waxy liquid which is visible under ultraviolet light.

ABOVE | Lizards shed or slough the outer epidermal layers of their skin in patches as they grow. The sloughed skin is transparent as the pigmented cells are located deeper in the dermis.

SENSE ORGANS

Sensory perception in lizards involves touch, taste and smell, hearing, and vision. As in other vertebrates, the stimulated sense organ communicates with the brain via the peripheral and central nervous systems, and the lizard responds accordingly. These sense organs are the means by which a lizard finds food and avoids noxious substances or non-foods, finds and recognizes a mate, monitors its position in the environment, and hopes to evade predation. While most of the sense organs of a lizard are similar to those found in mammals, some are more sensitive, certainly than those used by humans.

SCALE ORGANS

Many squamate reptiles possess mechanosensory scale organs, but these have not been much studied. One study of these organs in the desert-dwelling horned lizards (*Phrynosoma*) theorized that they may be pressure-sensitive and triggered by contact with objects in the lizard's path; detect vibrations caused by rainfall on the ground or the close proximity of potential predators; detect the precise location of attacking ants while the lizard is feeding; or even indicate the lizard's depth beneath the surface. These receptors will respond to touch and are probably also associated with processes such as basking to achieve optimum temperatures.

TONGUES AND NASAL AND VOMERONASAL ORGANS

Taste and smell are important senses, and while they operate as separate senses in humans they are strongly linked in squamate reptiles. The paired nasal organs are located in the nares (nostrils) and detect airborne scents, while the highly innervated vomeronasal organ (Jacobson's organ) is located in the roof of the mouth, where it can be accessed by the lizard's tongue bearing odorous particles. Some lizards also possess taste buds on the tips of their tongues and indulge in substrate or object licking. There is considerable variation in the shape of lizard tongues, in contrast to those of snakes, which are always long and forked. This is because lizards use their tongues for a wider variety of purposes, and not just as sense organs.

Monitor lizards (*Varanus*) and tegus (*Tupinambis*) possess forked tongues very similar to those of snakes, while the forks are present but shorter in the Gila Monster and beaded lizards (*Heloderma*). These forked tongues flicker in and out of the mouth, carrying particles to be deposited in the twin-channeled vomeronasal organs, the process occurring with greater frequency when the lizard is following a potential mate's or prey animal's scent trail, or when faced with a threat. Not all anguimorph lizards possess long forked tongues, those of the slow worms (*Anguis*) and Scheltopusik (*Pseudopus apodus*) being more rounded and notched at the tip. All the aforementioned lizards,

ABOVE LEFT | An iguana (*Ctenosaura* sp.) substrate-licks with the tip of its tongue.

ABOVE | A Komodo Dragon (*Varanus komodoensis*) has a long, forked tongue, one of its most important sense organs.

and the geckos, were at one time placed in the clade Scleroglossa, which means "hard tongues." Since scleroglossans use their powerful jaws and teeth for prey capture and manipulation, their tongues were free for chemosensory purposes.

In contrast to the scleroglossans, the Tuatara (*Sphenodon punctatus*) and the iguanians (agamas, chameleons, and iguanas) use their tongues to capture or pick up, and then manipulate, their prey, a process known as "lingual prehension", which means their tongues may have less sensory function. However, iguanas and agamas are often observed using the tips of their tongues to "lick" the substrate or an object and it is likely they are using their taste buds on the tongue to identify what is in front of them. While there is a small notch or cleft present in the tongue tip of most iguanians it is absent from the tongues of chameleons, and of the Tuatara.

Lizards also use their tongues for non-sensory and/or defensive purposes. Geckos clean the spectacles (brilles) covering their eyes by licking them with their tongues, while blue-tongued skinks (*Tiliqua*) extend their broad colorful tongues as part of their gaping threat display.

ABOVE | A chameleon's tongue may be longer than its own body, and this young Veiled Chameleon (*Chamaeleo calyptratus*) is putting it to good use catching insects.

RIGHT | The eye of a Tokay Gecko (*Gekko gecko*) with its vertically elliptical pupil, which opens widely to admit more light at night. Many diurnal geckos have round or almost round pupils.

RIGHT | The New Caledonian Crested Gecko (*Correlophis ciliatus*) lacks eyelids, in common with all geckos except the eublepharids. It possesses a transparent spectacle over the eye instead. It cannot blink, so cleans its eye with its tongue.

VISION AND EYES

The structure and function of lizard eyes varies greatly, from the vestigial scale-covered eyes of burrowing lizards, such as the dibamids, which may function only to warn when the lizard is exposed, to the eyes of active predatory lizards, which are capable of greater visual acuity than those of many mammals.

The typical reptile eyeball comprises a cornea and lens that focus light onto the retina, a photosensitive layer of cells at the back of the eye, the most sensitive part of which is the centrally positioned fovea. The iris of the eye surrounds an opening, the pupil, through which light passes. Embedded in the irises of many lizards are thin bony plates known as scleral ossicles, which protect the lens from external pressure and assist the ciliary muscles in changing the shape of the lens as the lizard focuses its vision. Snakes lack scleral ossicles or ciliary muscles, so focus is obtained by moving and distorting the lens forward or backward.

The retina contains two kinds of cells, known as rods and cones. Rods are sensitive in low light conditions but do not perceive color, while cones are sensitive to color but do not function well in low light conditions. The cones of diurnal lizards also contain colored lipid droplets that filter out particular wavelengths and focus the remaining light on the fovea, like colored filters on a camera lens. Diurnal lizards possess more cones than rods, with many taxa in the Iguania, Anguimorpha, Lacertidae, Pygpodidae, and Serpentes exhibiting cone-only retinas.

Nocturnal reptiles usually possess more rods than cones. Contrarily, the retinas of nocturnal geckos lack both rods and foveae, but possess more sensitive cones, plus rod-like visual cells that evolved from cones. The nocturnal Helmeted Gecko (*Tarentola chazaliae*) has specialized cones that are more sensitive to short-wavelength light than are those of diurnal lizards. Combined with its large eyes, vertically elliptical, cat-like pupils, which open very wide in the dark, and a short focal length to the retina, it can discern ultraviolet, blue, and green light in moonlight conditions. Some geckos have reverted back to diurnal activity, such as the Giant Madagascan Day Gecko (*Phelsuma grandis*), leading to the development of relatively small eyes, round pupils, small cones, and the re-evolution of rudimentary foveae.

Many lizards possess excellent visual acuity, and some have better color vision than humans; for example, the Helmeted Gecko and Desert Iguana (*Dipsosaurus dorsalis*), which can also see in ultraviolet. Vision is an important sense for lizards that rely on it to hunt, or to avoid being hunted, and is also important in behaviors such as courtship or territorial defense.

Most lizards possess eyelids and can blink, but all geckos (except the Eublepharidae) and a few genera from other families, including snake-eyed skinks (Scincidae: *Ablepharus*, *Cryptoblepharus*, and *Panaspis*), snake-eyed lizards (Lacertidae: *Ophisops*) and spectacled microteiids (Gymnophthalmidae: *Gymnophthalmus*), instead possess a clear spectacle (or "brille") similar to that of a snake. In contrast, many snake-like legless lizards, such as the slow worms (*Anguis*) and Scheltopusik (*Pseudopus apodus*), do possess eyelids and can blink.

Perhaps the strangest lizard eyes are the turret eyes of chameleons. Not only are these bulbous eyes almost entirely covered with skin so that only a small opening is available for vision, but they can also be used independently, so that a chameleon is able to look in front with one eye and behind with the other. Somehow the brain must deal with two conflicting images, but when the chameleon sights potential prey it will focus both eyes on its target and advance until it is close enough to use its long, projectable tongue to capture the meal.

ABOVE | A large Panther Chameleon (*Furcifer pardalis*) advances, using its independent turret-eyes to look in two directions at the same time.

THE PINEAL EYE

The Tuatara (*Sphenodon punctatus*) and some lizards, such as iguanians and monitor lizards, also possess a third eye, known as the pineal eye, parietal eye, or pineal organ. Located in the center of the top of the head, and connected to the brain, it is often clearly visible as a small, round structure. The pineal eye detects and monitors the presence or absence of light falling on the lizard's head, enabling it to optimize its temperature regulation through basking and informing it when night has fallen. The pineal gland, located in the epithalamus of the brain, then secretes the hormone melatonin, which controls sleep patterns. Stroking the pineal eye of an iguana will often cause it to close its eyes and seemingly go to sleep. It is important for maintaining a lizard's circadian (daily) rhythm and seasonal cycles.

ABOVE | The Elegant Earless Lizard (*Holbrookia elegans*) does possess tympana, but they are obscured by skin. It is therefore not deaf.

HEARING AND EARS

Squamate reptiles do not possess large, protruding ears like mammals, so there is no cartilaginous pinna to capture sound and direct it down the auditory canal to the tympanum (eardrum). Instead, the tympanum of lizards such as iguanas, many agamas, or monitor lizards is clearly visible, positioned at the outer end of the auditory canal, although a few species, including the majority of skinks, possess a countersunk tympanum with a shallow external auditory canal opening.

Not all lizards possess external ear openings. Rather like snakes, many fossorial or semi-fossorial reptiles have dispensed with external ears. The two most famous examples are the Tuatara (*Sphenodon punctatus*) and the secretive Borneo Earless Monitor (*Lanthanotus borneensis*), but earlessness is also found in the legless

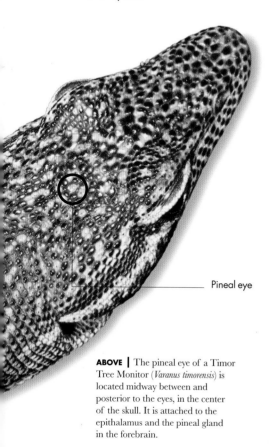

Pineal eye

ABOVE | The pineal eye of a Timor Tree Monitor (*Varanus timorensis*) is located midway between and posterior to the eyes, in the center of the skull. It is attached to the epithalamus and the pineal gland in the forebrain.

ABOVE | The Asian Water Monitor Lizard (*Varanus salvator*), on the left, has large and clearly visible tympana, whereas in the Tokay Gecko (*Gekko gecko*) on the right, the tympanum is located deeper in the slit-like auditory canal.

pygopodid worm-lizards (*Aprasia*), which contains 14 species, of which only one (*A. aurita*) possesses ear openings, and numerous skink genera. Some earless lizards have completely different lifestyles to these semi-fossorial species, being highly alert and diurnally active desert inhabitants, such as the Greater Earless Lizard (*Cophosaurus texanus*) and lesser earless lizards (*Holbrookia*), from the American deserts, and the Australian earless dragons (*Tympanocryptis*).

A lack of external ear openings does not mean the Tuatara, Borneo Earless Monitor, and other earless species are actually deaf. In the Tuatara the external ear opening is absent, there is no functional tympanum, and the middle ear is full of connective tissue (although this is not the case in earless squamates), but the inner ear is well developed, and Tuataras are believed to be able to perceive sound, especially as they vocalize themselves, with a series of croaks, during male combat, courtship, and when restrained. Certainly, it has been proposed that fossorial amphisbaenians and lizards may be able to detect low-frequency vibrations through their lower jaws or even though their lungs.

In the American earless lizards and Australian earless dragons the tympana are present but covered with skin, so they are not visible externally, but are capable of detecting airborne sounds. Snakes have gone one step further, having lost external ears, the tympanum, and the Eustachian tube which connects the middle ear to the throat, but they detect vibrations via their lower jawbones. The worm-lizards (*Aprasia*) have also lost their Eustachian tubes, and are the only Australian lizards to have done so.

LIFE IN EXTREME CONDITIONS

Lizards can be found in a wide range of environmental conditions, many of which seem inhospitable to life due to temperature, salinity, or aridity. These highly diverse reptiles have found a way not only to survive, but also to thrive in these hostile habitats.

COLD ENVIRONMENTS

Since reptiles are ectothermic they rely on the ambient climatic conditions to enable them to function, and that requires a daily period of basking. At higher elevations and higher latitudes, with shortened day lengths, this is a challenge, yet lizards are present in these cooler than optimum habitats. The Viviparous Lizard (*Zootoca vivipara*) occurs north of the Arctic Circle from Scandinavia and Siberia to Sakhalin Island, while the Strait of Magellan Lizard (*Liolaemus magellanicus*) is one of two species inhabiting Tierra del Fuego, the southernmost point of South America. These are the northernmost and southernmost lizards in the world. A candidate for the highest-occurring lizard in the world would be the Tuberculate Agama (*Laudakia tuberculata*), which occurs at elevations of up to 16,400 ft (5,000 m) in Kashmir.

The primary adaptations for life in cold environments are viviparity, whereby females become mobile incubators for their embryos, and melanism, the presence of dark pigment which promotes more rapid warming when basking. *Laudakia* is, however, oviparous.

RIGHT | A Viviparous Lizard (*Zootoca vivipara*) in snowy conditions.

LEFT & BELOW | The Shovel-snouted Lizard (*Meroles anchietae*) from the Namib Desert avoids overheating on hot sand by standing with its tail and alternate fore- and hind feet raised, before switching its feet over. During the hottest and driest part of the year this lizard obtains its water requirements from the seeds it eats, in the absence of any insect prey.

HOT ENVIRONMENTS

If a diurnally active lizard is standing on either hot rocks or hot sand for any length of time it is in danger of fatally overheating, but some species are extremely well adapted to such severe conditions. The Shovel-snouted Lizard (*Meroles anchietae*) inhabits the Namib Desert of southwestern Africa where the air temperature can easily top 110 °F (44 °C) and the sand can be too hot to stand on for very long. The lizard gets around this problem by running between shady desert shrubs at high speed (20 times its body length per second), but if it is forced to remain in the open it will do a little "dance," raising alternate fore- and hind limbs and its tail off the sand for a second or two, then swapping to the other feet. If it all becomes too much, it dives headfirst into the sand using its flattened, shovel-like snout, and disappears to where it is cooler.

ABOVE | The Galapagos Marine Iguana (*Amblyrhynchus cristatus*) feeds on seaweed on the ocean bed and rids itself of excess salt by snorting it from glands in its nostrils.

ARID ENVIRONMENTS

Another problem with living in hot climates is water conservation. Even in very hot and arid deserts there is often water available; the problem is knowing how to access it. Early morning fog is one source, especially in coastal deserts like the Namib, and the Shovel-snouted Lizard will drink water droplets from vegetation, taking in 15 percent of its body mass in water in three minutes.

For deserts located farther inland there is the possibility of collecting condensed early morning dew or even occasional rainfall. The Thorny Devil (*Moloch horridus*) of Australia and the American horned lizards (*Phrynosoma*) have very spiny backs, and not just to break up their outlines; these spinous scales collect rainwater and move it by capillary action toward the corners of the lizard's mouth so that it can obtain a drink, an efficient process known as rain-harvesting. The Thorny Devil can also draw up water from rain puddles by capillarity.

Other lizards have extremely efficient kidneys, so that little water is lost in their crystalline uric acid. And still others are able to obtain their entire water requirements from their diet, the Shovel-snouted Lizard acquiring most of its water from a diet of seeds when conditions are too hot for insect prey.

HIGHLY SALINE ENVIRONMENTS

Reptiles living in marine habitats, or those ingesting highly saline diets, must rid their bodies of excess salt. It is more efficient, and conserves more water, if they do this via salt excretory glands than through the kidneys as urine. Seasnakes pass crystalline salt from glands under the tongue into the ocean; the saltwater crocodile also excretes salt from glands on its tongue into the environment; and sea turtles dispose of excess salt via tears, from glands behind the eyes.

The Galapagos Marine Iguana (*Amblyrhynchus cristatus*) feeds entirely on seaweed, which has a high salt content. An overdose of salt can be rapidly fatal, and with its entire diet consisting of seaweed it overcomes the problem by snorting crystalline salt from specialized glands in its snout while it is basking.

Yet one lizard beats all others in its ability to survive high levels of salt in its body. The tiny Swollen-snouted Side-blotched Lizard (*Uta tumidarostra*) is endemic to Isla Coloradito, in the Gulf of California. There are no insects or spiders on this tiny island, so this little lizard has to feed on the only available invertebrate prey, sea slaters, in the intertidal zone. Although related to mainland side-blotched lizards that feed on ants, the island species is capable of withstanding levels of salinity that would be many times the lethal dose for other terrestrial vertebrates. Like Galapagos Marine Iguanas, they also rid themselves of crystalline salt by snorting it from their swollen snouts, and have specialized glands for this very purpose.

BELOW | The Swollen-snouted Side-blotched Lizard (*Uta tumidarostra*) feeds entirely on marine isopods in the intertidal zone on its remote island. It can withstand salt levels within its body that are many times those that would kill its mainland relatives. It disposes of excess salt by snorting it from salt excretory glands, crystalline salt being visible on its snout.

LOCOMOTION

Lizards are found in many diverse habitats, and each has its vertical and horizontal surfaces that they must traverse. Different substrates present different, or sometimes similar, challenges, and lizards have evolved and adapted to maximize their ability to move around in these diverse environments. The solutions they have adopted are as varied as the lizards themselves, so only a selection can be included here.

RUNNING

Most lizards possess four well-developed limbs, each with five digits. It has already been explained that the lizard's skeleton must be strong enough to raise its body off the ground and move it forward, and that lizards have a sprawling gait due to the positioning of the limbs outward and down, in contrast to the limbs of mammals. When a lizard runs, its body is thrown into a series of lateral S-shaped curves that are very evident in large species like monitor lizards, the long tail acting as a counterbalance. Terrestrial lizards often possess

ABOVE | The toes of the hind feet of the fringe-toed lizards (*Uma*), and some other genera, bear a fringe of projecting scales that enables the lizards to run on loose, shifting sand.

long, narrow fingers and toes, each terminating in a sharp claw to provide a purchase on the substrate as the lizard runs. A streamlined body is an advantage for rapid escape, but even fairly stout lizards, such as chuckwallas (*Sauromalus*) or dhab lizards (*Uromastyx*), can turn on the speed when required.

Some lizards have special adaptations for running on difficult substrates. The name "fringe-toed lizard" is applied to at least two unrelated genera, *Uma* (Phrynosomatidae) in the American deserts, and *Acanthodactylus* (Lacertidae) in North Africa, Iberia, and the Middle East. They have evolved the same trick for running across loose, shifting sand dunes. The toes of their hind feet are equipped with a fringe of protruding scales that spreads the weight of the lizard, preventing the toes from sinking into the loose sand when it kicks backward to move forward.

BELOW | The Australian Frilled Lizard (*Chlamydosaurus kingii*) can move surprisingly fast when it chooses to run bipedally.

ABOVE | A male Plumed Basilisk (*Basiliscus plumifrons*) runs across the water, its fringed toes spreading its weight and trapping air pockets so that it does not break the water surface.

The scales of the fringes may differ in their precise shape across all the lizards that possess them, but their purpose is the same, and they are also found in other dune-dwelling lizards such as the Persian Wonder Gecko (*Teratoscincus keyserlingii*), Rajasthan Toad-headed Agama (*Bufoniceps laungwalaensis*), and Namibian Desert Plated Lizard (*Gerrhosaurus skoogi*). Fringed toes in lizards provide a classic example of convergent evolution.

Lateral fringes on the toes of the hind feet are not confined to desert-dwelling lizards; they are also useful adornments for lizards living in aquatic conditions, who may need to flee a predator by taking to the water. The image of a Plumed Basilisk (*Basiliscus plumifrons*) running on two legs (bipedally) across the water is familiar to viewers of wildlife documentaries. The basilisk will manage at least a dozen paces, supported by the fringes of its toes, which spread its weight and trap air pockets, before exhaustion and the surface tension fail and it plunges in and is forced to swim. It is no surprise that in their native Catholic Central America the basilisks are called "Jesus Christ lizards," but they are not the only aquatic lizard that adopts this mode of locomotion. In northern South America the Mop-headed Lizard (*Uranoscodon superciliosus*) is also capable of sprinting across the water, while in Indonesia and the Philippines the relatively large, semi-aquatic sailfin lizards (*Hydrosaurus*) are also able to run short distances across the water surface.

Although possessing four limbs, the basilisks always run across water using a bipedal gait, and this technique also works on land, the Frilled Lizard (*Chlamydosaurus kingii*) of Australia providing an excellent example. When it feels threatened, it will spread its large, umbrella-like frill and open its mouth into a wide gape, the aim being to intimidate its enemy, but if this ploy fails the lizard will turn and sprint away bipedally to the nearest tree or hole, using the same gait as utilized by basilisks on the other side of the world.

LEFT | The Malagasy day geckos (*Phelsuma*) of Madagascar and the Indian Ocean islands, have dilated toes and are excellent climbers of trees and palms.

CLIMBING

Some lizards may have found fringed toes a solution to the problems of running in the horizontal plane where loose sand or water are involved, but this adaptation is also useful for moving in the vertical plane. South American keel-scaled teiids (*Kentropyx*) have fringed toes, and are primarily arboreal, climbing trees, shrubs, and even coarse grass tussocks. The basilisk's close but primarily arboreal relatives, the helmeted lizards (*Corytophanes*), also possess fringes along their toes, which presumably enhance their climbing abilities.

Lizards may climb rock faces, the walls of buildings, trees, and other vegetation. Sharp claws, such as those of the bent-toed geckos (*Cyrtodactylus* and *Cyrtopodian*) or iguanas, including green iguanas (*Iguana*) and the spinytail iguanas (*Ctenosaura*), are useful for climbing rapidly up rocks or trees, respectively, but the ultimate climbing toes are those with highly adapted "scansors," or toe pads. Again, no single taxon of lizards has a monopoly on these; they have evolved independently in several unrelated groups.

Expanded digital toe-pads or scansors are found in five of the seven gekkotan families, but even within those families, not all genera possess them. The scansors of arboreal geckos are dilated to provide the maximum surface area for adhesion when climbing, but that surface area is multiplied many times over by the structure of the scansors' underside. Close examination reveals rows and rows of plate-like structures known as lamellae, and these lamellae comprise millions of tiny branched setae—microscopic, hair-like projections with flattened, spatula-like tips. It is these tiny setae that provide the secret to how a gecko can run up a windowpane or acrobatically hang upside down from the ceiling. Tiny electrostatic forces exist between the tips of the setae and the substrate, and when these forces are multiplied by the millions of setae they are more than sufficient to maintain the gecko aloft. Controlled disengagement of the setae is necessary if the gecko is to continue moving, and this is achieved

RIGHT & BELOW | The left hind foot, from above and below (right) and first toe (below) of a Tokay Gecko (*Gekko gecko*), showing the lamellae, setae, and claws of the scansors.

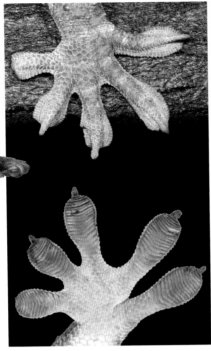

FAR RIGHT | The zygodactylous right forefoot of a Flap-necked Chameleon (*Chamaeleo dilepis*), from above and below, showing fused "mittens" of three inner and two outer toes.

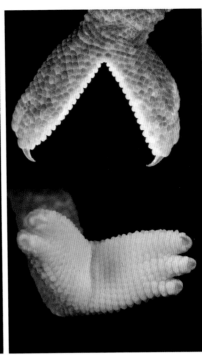

by rolling the toe forward so that the setae are rolled off the surface, before placing them down again in a more forward position.

Dilated scansors for climbing are not unique to the geckos; they are also found in the unrelated anoles (*Anolis*), and a few skinks such as the Papuan Green-blooded Skink (*Prasinohaema virens*). Adhesion is also not confined to the toes. The New Caledonian chameleon geckos (*Eurydactylodes*) and the Papuan green-blooded skinks (*Prasinohaema*) possess adhesive tail tips, enabling the tail to serve as a fifth limb when climbing. This ability to use the tail as an extra limb is also found in large lizards, such as the Solomon Islands Monkey-tailed Skink (*Corucia zebrata*) and the Madagascan Panther Chameleon (*Furcifer pardalis*), albeit without the adhesive tail tip.

This last species belongs to one of the most famous arboreal groups of lizards in the world, the chameleons, over 200 species of Afro-Madagascan lizards that share a raft of unique characteristics, the structure of their feet being just one example. The feet are "zygodactylous," with the five clawed toes of each foot divided into three and two toes and fused for combined strength. The forelimbs have two outer and three inner toes fused, and the hind feet have three outer and two inner toes fused, thereby providing equal numbers of outer and inner toes on each side. This arrangement enables a chameleon to grip even extremely slender twigs. While other lizards adopt a sprawling, serpentine gait, the laterally compressed body of a chameleon, with its legs extending almost vertically below, means that its center of gravity is located above its limbs and it is as stable as a tightrope walker when moving ponderously through the vegetation, with its prehensile tail providing an additional anchorage point.

ABOVE | The unsupported lateral webbing and skin flaps of Kuhl's Flying Gecko (*Gekko* (*Ptychozoon*) *kuhli*) enable it to parachute to safety.

GLIDING

The terms "flying lizard" and "flying snake" are commonly used for particular genera of Southeast Asian reptiles, but these terms are misleading. There are of course no extant flying non-avian reptiles, the pterosaurs having vanished at the end of the Cretaceous. Today the only flying animals are the insects, birds (which are strictly reptiles), and bats. Those squamates we refer to as "flying" are actually gliding, which is basically parachuting, making a controlled descent over a relatively short distance, rather than jumping or falling to the ground. Controlled it may be, but it is not flight, which requires that the animal be able to ascend as well as descend, and be able to sustain that process for a reasonable distance. There are two main genera of gliding lizards, both from Asia, but they have evolved different methods of gliding.

The 13 species of Southeast Asian flying geckos belong in the genus *Ptychozoon*, but its status has recently been reduced to a subgenus of genus *Gekko*. The flying geckos have a considerable amount of webbed skin, extending outward from the body, along the limbs and tail, and between all the toes. These webs or frills serve to disrupt the gecko's outline and reduce its shadow when it is hugging a branch, but they really come into their own when the gecko leaps from the branch and makes a controlled glide to another tree some distance away. None of these frills of skin are supported by specialized structures, so this escape mechanism is really parachuting, which is why some authors refer to them as "parachute geckos." The fairly heavily built turnip-tailed geckos (*Thecadactylus*) of Latin America also possess frills and webbing along their bodies and limbs and are reported to parachute, and it seems

ABOVE | The patagia of the Common Flying Dragon (*Draco volans*) are supported on long, extendable ribs.

ABOVE RIGHT | A Common Flying Dragon (*Draco volans*) will execute a slight upwards turn as it come to the end of its glide so that it lands on all four feet and facing up the next tree or palm trunk.

likely that this mechanism may be available to other frilled geckos.

The agamid genus *Draco* is large, with 40 species currently recognized from South and Southeast Asia. Known as flying dragons, they are gracile lizards with bulbous heads, long, slender limbs, and a thin tail. Along the flanks of these diminutive lizards runs a large fold of skin known as a patagium, which is supported by extremely elongate ribs that can be extended because lizards do not possess a sternum, which would make the rib cage more rigid and hold the ribs in place.

The patagia may be used in territorial displays, the color patterns being species-specific, but they are also a marvelous way to escape a potential predator, or a persistent suitor. The flying dragon simply leaps into space and plunges downward for a short distance before it spreads its patagia and alters its angle of descent to one that allows it to glide to safety on another tree, possibly as far as 98 ft (30 m) away. When coming in to land, the dragon will perform a small upturn of the body to enable it to land on all four legs and facing upward on the tree trunk. The forelimbs are spread wide during the glide, and it is likely this posture is necessary to enable the elongated ribs to expand forward and outward, while the tail may be used to alter the angle of the initial dive and the subsequent glide.

BURROWING

Fossorial lizards might be characterized by all or any of the following: elongate bodies, covered in smooth, often shiny scales; a lack of limbs, or forelimbs absent and hind limbs reduced to scaly flaps or tiny claws; a short tail; eyes absent, or reduced to pigmented cells under a translucent scale that distinguish only light from dark, or present but small, covered by a spectacle rather than movable eyelids; and a complete lack of any external ear openings. Other, more subtle characteristics could include a robust skull with fusion of the skull bones, and overlying fusion of the head scutes, and a snout specially adapted to enable the lizard to force itself forward through the substrate. Which structures are present or absent may differ depending on whether the lizard lives in an area of fine, shifting sand or a more resistant material such as soil, and how much time it spends above or below the surface.

The most fossorial squamate clade, apart from the Scolecophidia (blindsnakes and threadsnakes), is the Amphisbaenia, a group of almost 300 entirely fossorial tropical and subtropical burrowing reptiles. Most amphisbaenians or worm-lizards are legless, but they exhibit a wide variety of skull shapes, ranging from a rounded battering ram to a flattened shovel-snout, laterally compressed keel-snout, or even a pointed drilling snout, and they progress forward through the soil or sand by making side-to-side, up-and-down, or twisting motions of the head, dependent on their snout shape and the substrate in which they occur.

The most bizarre amphisbaenians are the ajolotes, or mole-lizards, of Mexico. The Five-fingered Ajolote (*Bipes biporus*) inhabits southern Baja California, where it constructs burrows in the sandy ground with entrances hidden under bushes or rocks. It has a strange appearance, being pink with a rounded head and an elongate body covered by annular scales, and while it has no hind limbs it does possess a pair of well-developed, mole-like forelimbs, which it puts to good use excavating its burrows.

ABOVE | The Arabian Sandfish (*Scincus mitranus*) has a shovel-snout and four powerful legs, allowing it to dive into a sand dune and "swim" rapidly downward to safety.

LEFT | The Five-fingered Ajolote or Mole worm-lizard (*Bipes biporus*) is a strange-looking sand dune inhabitant of the southern end of the Baja California Peninsula.

Numerous skinks are entirely fossorial, among them the southern African lance skinks (*Acontias* and *Typhlosaurus*) and the West African legless skinks (*Feylinia* and *Melanoseps*), which at first glance could be mistaken for blindsnakes. Similarly, the limbless sand microteiids (*Calyptommatus*) spend their lives within the sand dunes of the São Francisco River in northeast Brazil. However, not all burrowing lizards are legless, the Arabian Sandfish (*Scincus mitranus*) possessing four relatively well-developed limbs, but also a flattened shovel-snout. When it feels threatened it dives into the dune and "swims" downward rapidly through the sand, hence its common name.

REPRODUCTION

The biological aim of every individual organism is to pass on its genes to the next generation, and that usually means it must mate with a member of the opposite sex of its own species. Reproduction is also one of the means by which species adapt and evolve, becoming larger or smaller, evolving new traits or behaviors, adapting to changing conditions, and ultimately evolving into new species. But reproduction can be both a dangerous and costly business, for both males and females, and requires a considerable investment of time, effort, and resources to successfully accomplish it.

SEXUAL DIMORPHISM AND SEXUAL DICHROMATISM

When males and females exhibit different body shapes or proportions, or adult males possess imposing adornments such as crests, casques, or horns, this is termed sexual dimorphism. Many adult male lizards are physically larger than the females, although there are species where the female is the larger sex, possibly due to a requirement for her to carry more or larger eggs or offspring. Male Scheltopusiks (*Pseudopus apodus*) from southeastern Europe and Western Asia, and Guianan Caiman Lizards (*Dracaena guianensis*) from northern South America, possess much more robust heads than females, while adult male Veiled Chameleons (*Chamaeleo calyptratus*) exhibit larger crests than females or immature males.

When differences in coloration exist between the sexes this is termed sexual dichromatism, and there are numerous examples among the lizards. One such is the Rainbow Whiptail (*Cnemidophorus lemniscatus*). Males have a turquoise head, blue throat, bright green and yellow spotted flanks, a brown-striped dorsum, and a green tail, while females are less strikingly patterned, with a white-striped back and flanks and a yellow throat.

Lizards are very visually orientated, and they are also often highly visible reptiles, and many species take advantage of both sexual dimorphism and sexual dichromatism, as males seek to attract females and to deter rival males from encroaching on their territories or courting their females. That the male is the gaudy sex is common in the animal kingdom, and epitomized by peacocks and birds-of-paradise, but some male lizards are not far behind these birds when it comes to brilliant colors and stereotypical displays. Sexual dimorphism and sexual

LEFT | Male and female Veiled Chameleons (*Chamaeleo calyptratus*) can be distinguished by the differences in their head and casque sizes. This is a male.

ABOVE RIGHT | A brightly colored male Rainbow Whiptail (*Cnemidophorus lemniscatus*) attempting to subdue and mate with a less colorful, striped female of the same species.

dichromatism are extremely common across lizard families and may even be subtly present in secretive, cryptic, leaf litter-dwelling species.

COURTSHIP

Male lizards take every opportunity to display to females, but it is important they attract females of their own species. Where several closely related species occur in sympatry, such as the anoles (*Anolis*) in Latin America and the Caribbean, and the flying dragons (*Draco*) in Asia, the species can be distinguished, and sexual selection maintained, by the species-specific coloration and patterning of an extendable flap of skin under the throat, known as a dewlap. Dewlap coloration is so indicative of the species in locations with several otherwise identical species that scientists also rely on it as a primary characteristic for identification.

Courtship displays are not just about brightly colored banners. Many male lizards use a series of head-bobs, head-shakes, or push-ups to attract the female's attention, and again there is an analogy with the elaborate displays of birds-of-paradise or peacocks. The colors and the actions come as a complete package as the male lizard seeks to advertise his readiness and fitness to mate.

However, when a male lizard stands on an exposed vantage point and displays to attract females, or to defend his territory, he is taking a risk that he may also attract the wrong sort of attention, that of a predator. There is no reason to announce himself brightly when he is not courting, when he is sleeping, or outside of the breeding season, so there are times when survival demands that a male lizard be less conspicuous.

So dewlaps and other bright banners are folded away and the lizard becomes drab again. Male Common Agamas (*Agama agama*) are bright red or orange and blue or turquoise, but they only exhibit these colors when they have been basking; at night they are as drab as the gray-brown females. The male Sand Lizard (*Lacerta agilis*) is just one of numerous lizards that only adopts his most brilliant green robes during the breeding season.

TERRITORIALITY

Many of the displays used to attract females are also used to deter rival males. Male Green Iguanas (*Iguana iguana*) possess a large green dewlap, and orange-shouldered males in breeding condition use them to good effect, throwing the head back to almost vertical, extending the dewlap as much as possible, and going into an elaborate display of waving the dewlap and head-bobbing to exert dominance over a rival male. Often such displays diverge from those used to attract females when they start to include actual threats, such as inflating the body and curving it concavely to face the enemy, and standing tall on extended legs to enhance the impression of size, followed by tail-lashing, mouth-gaping, and, in some species, an audible hiss.

A fight is usually the last resort for a male lizard defending its territory, but if the threats are ignored by a similar-sized male, a fight is likely. Iguanas will whip their tails at their opponents, while monitor lizards stand on their hind feet, grasp their rival tightly with their forelimbs and attempt to wrestle him to the floor. For these large lizards, brute force is more important than the subtleties of color, but in smaller lizards both color and adornments play a part. The coloration of a male Jackson's Three-horned Chameleon (*Trioceros jacksonii*) will become more intense as he jousts with a rival male, using his triple horns, while male Asian Fan-throated Lizards (*Sitana ponticeriana*) indulge in extremely agile combat, launching themselves at one another like mountain goats, with their brilliant blue, black, and orange dewlaps flashing.

LEFT | With over 400 species of anole (*Anolis*) and up to eight to ten occurring in sympatry, the evolution of species-specific dewlap patterns and colors is essential to ensure males attract females of the same species. This species is the Yellow-tongued Anole (*Anolis scypheus*) from Ecuador.

TOP RIGHT | Male Jackson's Three-horned Chameleons (*Trioceros jacksonii*) joust in the trees.

RIGHT | A pair of Komodo Dragons (*Varanus komodoensis*) mating.

MATING

If rivals have retreated and a female has been successfully courted, the male lizard will attempt to mate with her. The male lies above the female and loops his tail under hers to raise her tail and make her cloaca accessible. The mating process is often accompanied by the male neck-biting the female, a behavior seen in a wide variety of species, from iguanas to Scheltopusiks. Male lizards, like snakes, possess a pair of hemipenes (sex organs), and the male will insert one hemipenis into the female and copulate, a process that takes from a few seconds in the Mallee Sand Dragon (*Ctenophorus fordi*) to 12 minutes in the Komodo Dragon (*Varanus komodoensis*).

The Tuatara (*Sphenodon punctatus*) demonstrates similar courtship and dominance displays to iguanas and also neck-bites the female during copulation, but actual copulation is different because, being a non-squamate, the male Tuatara lacks hemipenes and has to align his cloaca perfectly with the female's to allow sperm transfer. Tuataras may remain in copula for up to 90 minutes.

OVIPARITY

The ancestral reproductive strategy of modern reptiles is oviparity (egg-laying). It was the evolution of the amniotic (or cleidoic) egg, with an amniotic membrane to prevent it drying out, that enabled reptiles to conquer the land; unlike amphibians, they are not required to return to the water to reproduce. The amniotic egg contains a protective shell, two protective membranes (the chorion and amnion), a yolk to provide nutrition, and a sac-like allantois to store waste products and for respiration. All crocodilians, turtles and tortoises, and the Tuatara are oviparous, as are the majority of squamate reptiles (lizards, worm-lizards, and snakes). Of the 43 lizard and worm-lizard families, 26 are entirely oviparous, and a further 13 contain oviparous subfamilies, genera, or species (see table opposite).

The embryos of oviparous reptiles obtain all their nutrition from the egg yolk, a process known as lecithotrophy. Squamate reptile eggs are of two different types. The most common eggs are flexible, oval, oblong, or elongate, with leathery shells that comprise a thick shell membrane covered by a thin calcified outer shell layer. One of the most elongate eggs is that of the Mbanja Worm-lizard (*Chirindia ewerbecki*), which may be 15 times as long as wide. The second type of egg is found in the gekkotan families Gekkonidae, Sphaerodactylidae, and Phyllodactylidae. The eggs of these geckos are composed of a thin shell membrane and a thick, calcareous shell, resulting in eggs that are rigid and often spherical in shape. The geckos that lay rigid-shelled eggs are divided into the "gluers" or "non-gluers," depending on whether or not their eggs are adhesive. While the leathery-shelled eggs and the eggs of the non-gluers are usually hidden under rocks or logs on the ground, the eggs of the gluers are laid aloft, using their adhesive surface to stick them under tree bark, in rocky crevices, in buildings, or on the undersides of leaves.

Most geckos and gymnophthalmids lay pairs of eggs, but smaller species, such as the least geckos (*Sphaerodactylus*), bachias (*Bachia*), and anoles (*Anolis*), produce only single eggs, and even some large species, such as the Giant Turnip-tailed Gecko (*Thecadactylus rapicauda*), produce only a single large egg. The female Graceful Crocodile Skink (*Tribolonotus gracilis*) lays a single striated,

RIGHT | A female Graceful Crocodile Skink (*Tribolonotus gracilis*) lays only a single egg equal to eight to nine percent of her body weight.

REPRODUCTIVE STRATEGIES OF LIZARDS AND WORM-LIZARDS

Blue = oviparous (lay eggs); Yellow = viviparous (bear live young); Green = oviparous & viviparous

INFRAORDER	FAMILY	SUBFAMILY	INFRAORDER	FAMILY	SUBFAMILY
Dibamia	Dibamidae		Amphisbaenia	Amphisbaenidae	
Gekkota	Carphodactylidae			Blanidae	
	Diplodactylidae			Bipedidae	
	Eublepharidae	Aeluroscalabotinae		Cadeidae	
		Eublepharinae		Rhineuridae	
	Gekkonidae			Trogonophidae	
	Phyllodactylidae		Iguania	Agamidae	Agaminae
	Pygopodidae	Lialisinae			Amphibolurinae
		Pygopodinae			Draconinae
	Sphaerodactylidae				Hydrosaurinae
Scincomorpha	Cordylidae	Cordylinae			Leiolepidinae
		Platysaurinae			Uromastycinae
	Gerrhosauridae	Gerrhosaurinae		Chamaeleonidae	
		Zonosaurinae		Leiosauridae	Enyaliinae
	Xantusiidae	Cricosaurinae			Leiosaurinae
		Lepidophyminae		Opluridae	
		Xantusinae		Liolaemidae	
	Scincidae	Acontiinae		Corytophanidae	
		Egerniinae		Crotaphytidae	
		Eugongylinae		Dactyloidae	
		Lygosominae		Iguanidae	
		Mabuyinae		Hoplocercidae	
		Scincinae		Leiocephalidae	
		Sphenomorphinae		Phrynosomatidae	Phrynosomatinae
Lacertoidea	Teiidae	Callopistinae			Sceloporinae
		Teiinae		Polychrotidae	
		Tupinambinae		Tropiduridae	
	Alopoglossidae		Anguimorpha	Anguidae	Anguinae
	Gymnophthalmidae	Cercosaurinae			Anniellinae
		Gymnophthalminae			Gerrhonotinae
		Riolaminae		Diploglossidae	
		Rhachisaurinae		Xenosauridae	
	Lacertidae	Gallotiinae		Helodermatidae	
		Lacertinae		Shinisauridae	
		Eremiadinae		Lanthanotidae	
				Varanidae	

leathery-shelled egg equivalent to eight or nine percent of her body weight and 26–32 percent of her SVL. Much larger clutches are found in larger lizards, such as the Nile Monitor (*Varanus niloticus*), which may lay up to 60 eggs, and even some small lizards, such as the horned lizards (*Phrynosoma*), with clutches of 12–45 very small eggs. Although many geckos and anoles can produce several small clutches a year (multi-clutching), those living in cooler southern regions, such as Darwin's Marked Gecko (*Homonota darwinii*) in Argentina, produce only one clutch every other year.

Females of many lizard species, such as monitor lizards or iguanas, excavate deep nests in the earth, or in sandy banks, while females of smaller species deposit their eggs in tree holes or damp leaf litter, and leave them to the vagaries of the weather. It is important that the nests are not flooded or the embryos will drown, so nest sites are chosen accordingly. Nest guarding is relatively uncommon in lizards, but a few species do remain with their eggs, notably the American Five-lined Skink (*Plestiodon fasciatus*), which will even devour any suboptimal eggs liable to become infected with fungi and threaten the rest of the clutch, while the Asian Long-tailed Sun Skink (*Eutropis longicauda*) will actively defend its nest against egg-eating snakes.

Lizards hatch by breaking out of their eggs, after "pipping" (cracking) them with a tiny egg tooth on the front of the upper lip. The time taken to hatch varies greatly. For small, vulnerable species it may be very rapid, just a few seconds, but for other species it may be quite labored, the hatchling resting for hours while it absorbs the remaining egg yolk.

BELOW | A female Five-lined Skink (*Plestiodon fasciatus*) will remain with her eggs and guard them.

RIGHT | A neonate Viviparous Lizard (*Zootoca vivipara*) breaks through the membrane and takes its first breath, its yolk sac visible.

VIVIPARITY

Viviparity, or live-bearing, has evolved at least 100 times in squamate reptiles. Around 20 percent of all lizards are viviparous, but there is not just one single mode of viviparity, so the term "ovoviviparity," whereby eggs hatch within the female just prior to birth, and therefore appear as live-born young, has fallen out of favor. Of the 43 lizard and worm-lizard families, three are entirely viviparous (Xantusiidae, Xenosauridae, and Shinisauridae), while 13 families contain viviparous subfamilies, genera, or species (see table on page 53).

Viviparity is thought to have arisen through egg retention by the female, possibly in response to inclement conditions, the "cold-climate hypothesis," enabling her to bask in the sun to augment development of her embryos. It also removes the risk of eggs being found and devoured by egg-eating snakes, destroyed by fungal infections, or flooded by rising water levels. But viviparity also has its negative consequences, placing an enormous burden on the female, who may be unable to feed during the gestation period, and, being larger and slower-moving, is potentially more vulnerable to predation.

One of the best-known live-bearing species is the Viviparous Lizard (*Zootoca vivipara*) of Europe and Asia, the only truly live-bearing species in the Lacertinae, whose viviparity enables it to protect its offspring from the cold and survive north of the Arctic Circle in Scandinavia and Siberia. However, its population in northeastern Spain and southwestern France is oviparous. The Sand Lizard (*Lacerta agilis*) lays eggs, but not until they are only a few days from hatching, having retained them in her oviducts throughout the gestation period.

ABOVE | The female Solomons Monkey-tail Skink (*Corucia zebrata*) usually only produces one large neonate which she then protects. The neonate will eat its mother's feces to obtain the microbial gut flora required to digest vegetation.

Eggs retained in the oviducts for any period of time experience a reduction in the thickness of their shells, resulting in the thin membranes that surround newly born neonates. If the process of shell development is halted completely there is the possibility for embryos to obtain their calcium and nutrients directly from the female, across the fetal membranes, in a process known as matrotrophy, as documented in the Tussock Cool-skink (*Pseudemoia entrecasteauxii*) from New Zealand. Although viviparity and matrotrophy were thought to have evolved together, this may not be the case, because there are viviparous lizards that retain lecithotrophy, obtaining all their nutrition from the egg yolk, like oviparous embryos, for example, the Pygmy Short-horned Lizard (*Phrynosoma douglasii*). At the other end of the scale, Heath's Skink (*Brasiliscincus heathi*), from arid northeastern Brazil, has developed placentotrophy, where the nutrients pass from mother to embryo via a placenta containing the parental and fetal blood vessels, similar to the placentas of eutherian mammals.

As with clutch sizes, litter sizes vary greatly across the lizards. The Monkey-tailed Skink (*Corucia zebrata*), a large species from the Solomon Islands, usually gives birth to a single large neonate, and if she produces twins they are substantially reduced in size and vigor. Contrastingly, the Eastern Blue-tongued Skink (*Tiliqua scincoides*) may produce up to 19 small neonates.

A few lizards care for their offspring, mostly members of the Egerniinae (social skinks), such as the Monkey-tailed Skink with its single neonate. The Black Rock Skink (*Egernia saxatilis*) lives in small family groups where the neonates obtain some protection by remaining with their mother in her burrow. However, most neonate lizards are independent from the moment they are born.

PARTHENOGENESIS

Parthenogenesis means virgin birth, reproduction without the involvement of a male. Obligate parthenogenetic species are those which always reproduce parthenogenetically, and there are no known males. Facultative parthenogenesis occurs when females of a normally sexual species produce offspring in the absence of any males. This latter form of parthenogenesis would seem to be a last chance form of reproduction for a normally sexual female. Among squamates, it has so far been reported in boas, pythons, filesnakes, gartersnakes, rattlesnakes, and even the Komodo Dragon (*Varanus komodoensis*). It is most often reported in captivity.

There is only one known obligate parthenogenetic snake species, the tiny Brahminy Blindsnake, or Flowerpot Snake (*Indotyphlops braminus*), which has been transported all around the world in the root balls of commercial crops and pot plants, easily founding new populations since it does not require a mate to start a new colony. However, there are at least 25 obligate parthenogenic lizard species. Many are created by hybridization events between two sexual species, resulting in entirely new parthenogenetic species, whereas others comprise entirely parthenogenetic populations of otherwise sexual species.

At least four species of Indo-Pacific gecko, and one Australian gecko, possess parthenogenetic populations: the Indo-Pacific House Gecko (*Hemidactylus garnotii*), Indo-Pacific Tree Gecko (*Hemiphyllodactylus typus*), Mourning Gecko (*Lepidodactylus lugubris*), Pelagic Gecko (*Nactus pelagicus*), and Bynoe's Prickly Gecko (*Heteronotia binoei*). The first three are excellent colonizers that have become distributed widely on islands across the Pacific and Indian Oceans.

In Central and South America, at least nine species of southwestern US and Mexican whiptails (*Aspidoscelis*) are parthenogenetic, as are Underwood's Spectacled Microteiid (*Gymnophthalmus underwoodi*), the Carinate Litter Microteiid (*Loxopholis percarinatum*), and Von Borcke's Keel-scaled Teiid (*Kentropyx borckiana*), all from northern South America.

In the Caucasus and Turkey at least five (from 34) species of rock lizards in the lacertid genus *Darevskia* are parthenogenetic, while in Southeast Asia the Thai Butterfly Lizard (*Leiolepis triploida*) is also a parthenogenetic hybrid. It is likely more parthenogenetic lizards will be identified in future.

RIGHT | The Indo-Pacific House Gecko (*Hemidactylus garnotii*) is parthenogenetic, existing as female-only populations, which makes it a good colonizer, but not a good competitor when the sexual Asian House Gecko (*H. frenatus*) arrives. The swollen white area in the throat is stored calcium for eggshell production.

DIET

Squamate reptiles come in all sizes and they view a wide array of living organisms, from ants to antelopes, as potential prey, but while the snakes are consummate carnivores, the lizards have also ventured into herbivory, feeding on a wide range of vegetation, from desert cacti to seaweed. Some diets require specialized teeth, and lizards demonstrate a much wider range of dentition than do snakes (see page 24).

HERBIVOROUS LIZARDS

The familiar image of a captive Green Iguana (*Iguana iguana*) munching its way through a bowl of greens may have given rise to the idea that herbivory defines the lifestyles of most lizards, but this is not actually the case. Vegetation does provide the basis of the diet for a number of large lizards, but while many species do occasionally eat small quantities of vegetation, relatively few are obligate herbivores. An obligate herbivore requires a specialized digestive system and a microbial gut flora that can digest cellulose, so facultative herbivores or omnivores are more likely to feed on easily digestible plant parts, such as fruit and flowers. Seasonal availability will also play a part in when lizards can eat vegetation, and even obligate herbivores will occasionally eat insects, carrion, or smaller vertebrates.

HERBIVORY IN THE IGUANIA

The Iguanidae is an almost entirely herbivorous family, with a preference for the color yellow. Green iguanas (*Iguana*), spinytail iguanas (*Ctenosaura*), the chuckwallas (*Sauromalus*), and the Fijian iguanas (*Brachylophus*) feed on the flowers, fruit, and leaves of a diverse variety of perennial plants, being able to avoid toxic plants by licking them for chemical cues. Even the smallest species, the terrestrial desert iguanas (*Dipsosaurus*), are primarily herbivorous, while species living on arid islands, such as the Galapagos land iguanas (*Conolophus*) or West Indian iguanas (*Cyclura*), feed on cacti pads, flowers, and fruits. The Galapagos Marine Iguana (*Amblyrhynchus cristatus*) is unique in that it dives into the cold Humboldt Current and swims to depths of 33 ft (10 m) to feed on brown and red algae (seaweed). When it returns to the black lava rocks it must bask to raise its body temperature to that required for digestion, a process that is accelerated by the iguana's dark pigmentation.

The exceptions to the total herbivory of iguanids are the Black Spinytail Iguana (*Ctenosaura similis*) and Western Spinytail Iguana (*C. pectinata*), which feed on arthropods as juveniles and shift to a fully herbivorous diet when they become adults.

LEFT | The Galapagos land iguanas (*Conolophus*) readily feed on cactus plants, spines and all.

ABOVE | The Ornate Day Gecko (*Phelsuma ornata*) lapping nectar from flower heads.

Wild green iguanas do not go through such a shift; juveniles are as vegetarian as adults, despite the fact that they will eat invertebrates under captive conditions.

Herbivory, or at least significantly plant-based omnivory, is also found in other iguanians and is not restricted to large-bodied species living in tropical environments. *Liolaemus* (Liolaemidae) is a genus comprising over 270 species of small Andean and Patagonian lizards sometimes known as swifts. Herbivory has evolved more times independently in this genus than in any other, even in species found in cold habitats, such as the Los Nascimentos Swift (*L. poecilochromus*), living at 11,500–13,500 ft (3,500–4,130 m) in the southern Andes, and the Strait of Magellan Lizard (*L. magellanicus*), on Tierra del Fuego, the southernmost lizard species in the world. Species in the related Andean lizard genus *Phymaturus* are also reportedly herbivorous.

In the Agamidae, herbivory is primarily confined to three small subfamilies. The sailfin lizards (*Hydrosaurus*) in the Moluccas, Indonesia, and the Philippines are primarily folivorous (leaf-eating) or frugivorous (fruit-eating), the butterfly lizards (*Leiolepis*) are also primarily herbivorous, while the dhab lizards and mastigures (*Uromastyx* and *Saara*) undergo a shift from an invertebrate juvenile diet to a folivorous adult diet. A number of other agamids also eat some vegetation, such as the Bearded Dragon (*Pogona barbata*).

HERBIVORY IN THE GEKKOTA

Geckos would seem unlikely vegetarians, but a few of the diplodactylids and gekkonids are facultative herbivores. The Raukawa Gecko (*Woodworthia maculata*) from New Zealand, the Gargoyle Gecko (*Rhacodactylus auriculatus*) from New Caledonia, and the Mauritius Blue-tailed Day Gecko (*Phelsuma cepediana*) feed on fruit and nectar, the last species being the sole pollinator and seed disperser for the rare montane plant *Roussea simplex* (Rousseaceae).

HERBIVORY IN THE SCINCOMORPHA

There are over 1,700 species of skinks and most of them feed on invertebrates, but the Scincidae does include a few herbivorous species, mostly larger species in the subfamily Egerniinae. The most famous is the Monkey-tailed Skink (*Corucia zebrata*) from the Solomon Islands, an obligate herbivore that feeds on a variety of plants but shows a preference for the aroid *Epipremnum* (Araceae), despite the fact these plants contain toxins. Monkey-tailed Skinks are also coprophagic; they eat each other's feces as a means of topping up the necessary microbial gut flora required for herbivory. One of the first acts of a neonate skink is to ingest the feces of its mother, because it is not born with the necessary gut flora. Some of the larger blue-tongued skinks (*Tiliqua*) are facultative herbivores, the most herbivorous species being the slow-moving Shingleback (*T. rugosa*), while Cunningham's Skink (*Egernia cunninghami*) is also primarily herbivorous.

Some African flat lizards, such as the Cape Flat Lizard (*Platysaurus capensis*), supplement their invertebrate diet with leaves, petals, and seeds. The Desert Plated Lizard (*Gerrhosaurus skoogi*) eats grass, seeds, and dried plant material in the Namibian sand dunes, while the Central American cave-dwelling Smith's Tropical Night Lizard (*Lepidophyma smithii*) feeds almost entirely on figs that drop into its cave.

HERBIVORY IN THE LACERTOIDEA

The most herbivorous members of the Lacertidae belong to the Gallotinae, including the Tenerife

RIGHT | Both the Thorny Devil (*Moloch horridus*), pictured, and the Regal Horned Lizard (*Phrynosoma solare*) feed almost entirely on ants.

Gallotia (*Gallotia galloti*), a large lizard that feeds primarily on fruit throughout much of the year, only resorting to arthropods during the winter. There are also several primarily herbivorous species among the teiid lizards. The Aruba Whiptail (*Cnemidophorus arubensis*) and Curaçao Whiptail (*C. murinus*) are island species that primarily feed on vegetation, but the widespread mainland Rainbow Whiptail (*C. lemniscatus*) is also herbivorous. Two of the three related desert teiids of South America are herbivorous: the Coastal Desert Teiid (*Dicrodon guttulatum*) eats primarily seeds, while the Libertad Desert Teiid (*D. holmbergi*) feeds on leaves and grasses.

HERBIVORY IN THE ANGUIMORPHA

Probably the least likely herbivorous lizards are the monitor lizards (*Varanus*), but the three species in subgenus *Philippinosaurus* are primarily herbivorous. Gray's Monitor (*V. olivaceus*), the Panay Monitor (*V. mabitang*), and the Sierra Madre Monitor (*V. bitatawa*) are arboreal rainforest inhabitants that exhibit a preference for ripe fruit from screw pines, figs, and kamuling, although they also supplement their diets with insects, arachnids, crustaceans, and mollusks.

INSECTIVOROUS OR CARNIVOROUS?

Terminology can be misleading. An animal that feeds on invertebrates is described as "insectivorous," but strictly that term should only be applied to those species that feed on insects, not spiders, woodlice, worms, or snails. Similarly, an insectivore is often seen as being different from

RIGHT | The Guianan Caiman Lizard (*Dracaena guianensis*) uses its powerful jaws and robust teeth to crush large snails.

BELOW | A Southern Alligator Lizard (*Elgaria multicarinatus*) crunching a large cricket.

a carnivore, the latter term being reserved for species that prey on vertebrates that comprise "meat," but an insectivore is biologically just a carnivore that eats small invertebrates, as exemplified by "carnivorous plants" which feed on flies. Therefore, the term "insectivore" is avoided here.

INVERTEBRATE CARNIVORES

The vast majority of the world's more than 7,000 species of lizards and worm-lizards are small, less than 6 in (150 mm) SVL, and only a relative few species feed on vegetation, so the majority must eat other animals. Being small lizards, this means fairly small animals, the legions of invertebrates that inhabit the rainforests, deserts, and grasslands where lizard diversity is at its greatest. Many of the invertebrates taken as prey will be insects—ants and termites and their larvae, springtails and silverfish, beetles and their larvae, butterfly and moth larvae (caterpillars), flies and their larvae, grasshoppers, locusts, katydids, mantids, and true bugs—but also included in the diverse diets of these lizards will be spiders, solifugids, and scorpions (all arachnids), centipedes (myriapods), woodlice and slaters (crustaceans), slugs and snails (mollusks), and earthworms (annelids).

Most small lizards are generalist or opportunist invertebrate feeders, with extremely catholic (euryphagous) diets, and will occasionally also consume vegetation, feces, carrion, or smaller lizards. But there are also a great many interesting specialist (stenophagous) invertebrate feeders.

MYRMECOPHAGOUS LIZARDS

Myrmecophagy means feeding on ants. Ants can occur in very large numbers, and they do not run from a predator, so a lizard feeding on ants can simply stand there and eat them all day, which is what it must do to obtain sufficient nutrition from such a diet. In the Sonoran Desert the Regal Horned Lizard (*Phrynosoma solare*) feeds almost entirely on ants, using its tongue and a bobbing motion of the head to pick them up. This species and the Desert Horned Lizard (*P. platyrhinos*) spend hours in the sun, exposed to predators and relying on their cryptic pastel coloration and spiny backs to avoid being predated. Where more than one species of horned lizard occur in sympatry, they feed on different species of ants, which may include highly venomous harvester ants (*Pogonomyrmex*). This sharing of available resources to avoid competition is called "resource partitioning." On the other side of the world, in the central Australian deserts, the Thorny Devil (*Moloch horridus*), a spiny agamid lizard, specializes in feeding on rainbow ants (*Iridomyrmex*), which it will eat at a rate of 45 ants a minute, up to 1,350 over an entire day. It is not known how these lizards cope with the ants' formic acid, but the stings do not seem to affect them, and horned lizards have also been known to eat bees.

Phrynosoma and *Moloch* provide an interesting example of convergent evolution, but they are not alone in feeding on ants. Tree-dwelling species, such as the Amazonian treerunners (*Plica*) and Asian flying dragons (*Draco*), feed largely on ants on the tree trunks. Other myrmecophagous species include the side-blotched lizards (*Uta*), the Chilean Mountain Lizard (*Liolaemus monticola*), the Desert Spiny Lizard (*Sceloporus magister*), and Amazonian thornytail lizards (*Uracentron*).

MOLLUSK-EATING LIZARDS

There is an entire guild of snakes that specializes in eating slugs and snails, but a molluskophagous diet is less common among lizards. This may be because the snakes that feed on mollusks, colloquially known as "goo-eaters," possess specialized oral glands to deal with the mucus

RIGHT | The Panther Chameleon (*Furcifer pardalis*) uses its zygodactylus hindfeet and prehensile tail to anchor itself firmly as it reaches out and propels its long sticky tongue to capture a cricket.

produced by their prey. The two sides of a snake's lower jaw are independent, so snail-eating snakes have evolved a technique whereby they insert one side of the lower jaw past the operculum of a snail, snag its body with a recurved tooth, and withdraw the prize.

The lower jaws of lizards are fused at the chin and operate as a single unit, most do not have snake-like teeth, and they probably do not possess specialized anti-mucus glands, so mollusks may prove a more difficult prey type for lizards. Yet the Guianan Caiman Lizard (*Dracaena guianensis*), a large, green South American lizard that is both arboreal and semi-aquatic, has evolved an efficient way of feeding on large aquatic Amazonian snails. Its most striking characteristics are its large head and powerful jaws, which are filled with robust teeth. The snail is picked up in the jaws and then rolled to the back of the mouth where it is crushed by the teeth like a nutcracker, the pieces of shell being flicked from the mouth by the tongue while the soft parts are swallowed. Other lizards that also occasionally crush and eat snails include the Nile Monitor (*Varanus niloticus*), which feeds on anything it can fit into its mouth, and the Scheltopusik (*Pseudopus apodus*), another powerful-jawed lizard capable of crushing snail shells to access their contents.

OTHER LIZARDS WITH SPECIALIZED DIETS

As already explained (page 39), the Swollen-snouted Side-blotched Lizard (*Uta tumidarostra*) dwells on a small island where the only available invertebrate prey comprises sea slaters in the intertidal zone. This isolated population of a widely distributed mainland species complex (*U. stansburiana*) can survive extremely high levels of salt in its diet.

The largest amphisbaenian, the Red Worm-lizard (*Amphisbaena alba*), inhabits Amazonia, where it follows the pheromonal

trails of leaf-cutter ants back to their nests. Once inside the nest it seems immune to attack as it seeks out the ants' fungal garden middens, and here it feeds primarily on the larvae of beetles that are themselves feeding on the vegetational debris.

Among the most specialized invertebrate-hunters are the chameleons. With their cryptic patterning, slow, swaying gait, and independently functioning turret eyes, they are superbly adapted for hunting large, alert, diurnal insects. When potential prey is sighted, the chameleon brings both of its eyes to bear and approaches slowly, swaying from side to side as it determines the distance to its target, mouth slightly open, and adhesive tongue visible and ready for action. When the chameleon is satisfied that it is within range it will suddenly propel its long tongue forward, the swollen adhesive tip contacting and adhering to the insect, which is yanked back into the chameleon's mouth and crushed with its teeth. Large Oustalet's Chameleons (*Furcifer oustaleti*) can project their tongues to capture prey 13¾ in (35 cm) away at an impressive speed of 19 ft/sec (5.8 m/sec, or almost 13 mph/21 kph), but it is believed that smaller species may be even faster.

VERTEBRATE CARNIVORES

Many moderate-sized lizards will engage in saurophagy (lizard eating), the collared lizards (*Crotaphytus*), leopard lizards (*Gambelia*), Tokay Gecko (*Gekko gecko*), Brown Sheen Skink (*Eugongylus rufescens*), and Namaqua Chameleon (*Chamaeleo namaquensis*) being diverse examples. Many will even devour smaller members of their own species. Among lizard specialists, Burton's Snake-lizard (*Lialis burtonis*), from Australia and southern New Guinea, has a long snout capable of great flexibility, which enables it to capture, kill, and swallow relatively large skinks in relation to its own body size (see page 116).

Most monitor lizards (*Varanus*) are carnivores, feeding on a variety of size classes of invertebrates and vertebrates (but see also herbivorous lizards, page 58), and most will readily eat carrion. Monitor lizards are found in Africa, Asia, and Australasia, and their niche is filled in South America by the heavy-bodied tegus (*Tupinambis*). Juvenile Komodo Dragons (*Varanus komodoensis*)

spend the first two years of life living in the canopy, feeding on large invertebrates and small lizards, safe from the attentions of potentially cannibalistic adults, including their own parents. The adult dragons are well known for their ability to ambush large prey such as goats or deer. It was thought the carrion-feeding dragons possessed a bacterial broth in their mouths that caused rapid and unstoppable septicemia, via the wounds they inflicted during their ambushing attack, but now it is believed there is also venom involved. There is only one case of a human being eaten by a Komodo Dragon and the details are sketchy, but one must wonder what size of prey its larger ancestor, the Megalania (*V. priscus*), was capable of running down and overpowering, and whether it included early humans.

BELOW | Three adult Komodo Dragons (*Varanus komodoensis*) can polish off an entire goat carcass in 20 minutes.

DEFENSE

Lizards have many enemies, both large and small, and they have evolved a diverse array of defensive strategies to avoid being predated. Defenses range from making themselves seemingly invisible to predators, to the complete opposite, boldly facing them down with a series of intimidating tactics designed to make the perceived enemy change its mind and leave the lizard in peace. In extremis, many lizards are even prepared to sacrifice a piece of themselves in order to escape. And some species adopt even more bizarre tactics to stay alive.

PREDATORS AND ENEMIES OF LIZARDS

In a world of carnivores, lizards are assailed on all sides by animals that may eat them. Most lizards are small, and they occasionally feature in the diets of large invertebrates. Spiders ranging from the Australian Redback Spider (*Latrodectus hasselti*) and European and American widow spiders, to large wolf spiders, tarantulas, and bird-eating spiders, will predate small geckos and skinks. Scorpions from at least three families (Scorpionidae, Buthidae, and Vaejovidae), the misnamed sun-spiders or camel-spiders (Solpugidae), and vinegaroons or whip scorpions (Uropygi) are also occasional predators, while ectoparasitic mites and ticks make their homes on the lizards themselves, feeding on skin or drinking blood. Lizards of several families possess "axillary mite pockets," like armpits, to accommodate these parasites, which congregate in the pockets, thereby reducing their presence on the limbs—which would affect running or abrade the lizard's skin—or in the lizard's ears. Predatory insects operating en masse, such as driver ants, or individually, such as large ground beetles and praying mantises, will eat lizards, while giant centipedes (Scolpendromorpha) are also significant predators.

If parasites are to be considered predators, then infestations of roundworms and tapeworms also pose a threat to a lizard's health and well-being, and while there are five human malarias, there are more than 90 species of malarial parasites that infect lizards, many of them species-specific.

There are numerous saurophagous (lizard-eating) vertebrates, and the most obvious are snakes. The highly arboreal, long-headed vinesnakes (*Ahaetulla* in Asia, *Oxybelis* in America,

LEFT | A female Banded-legged Golden Orb Weaving Spider (*Nephila senegalensis*) with her prey, an African dwarf gecko (*Lygodactylus* sp.).

ABOVE RIGHT | A Burmese shrike (*Lanius collurioides*) with her prey, a butterfly and a lizard, probably a lacertid grass lizard of genus *Takydromus*.

and *Thelotornis* in Africa) possess excellent vision that enables them to locate even the most camouflaged of lizards, while the Boomslang (*Dispholidus typus*) of Africa frequently preys on chameleons. There are also numerous saurophagous terrestrial snakes, such as the Central American Lizard-killer (*Coniophanes lineatus*), which specializes in ameivas and whiptail lizards, or the Lizard Egg-eating Snake (*Drepanoides anomalus*) from the western Amazon, which eats the eggs of several lizard species. Many desert-dwelling snakes, such as the Namib Sidewinding Adder (*Bitis peringueyi*), also prey primarily on lizards, which are captured from ambush. The invasive Common Island Wolfsnake (*Lycodon capucinus*) may be one of the factors behind the extinction or extirpation of the Réunion Skink (*Gongylomorphus borbonicus*) from Réunion in the Indian Ocean, while the Brown Treesnake (*Boiga irregularis*) has impacted lizard as well as bird populations of Guam. Voracious lizards, such as the collared lizards (*Crotaphytus*) and Tokay Gecko (*Gekko gecko*), predate smaller lizards, and cannibalism, especially of hatchlings, is known from species ranging from house geckos (*Hemidactylus*) up to the Komodo Dragon (*Varanus komodoensis*).

Many mammals and birds also prey on lizards. Small carnivores, such as the Meerkat (*Suricata suricatta*) and other mongooses, are avid predators of lizards, while feral, introduced, and invasive cats, foxes, dogs, and rats pose serious threats to the survival of endangered species, especially on islands in the Galapagos and Caribbean, or the islands off the coast of New Zealand, which are inhabited by the Tuatara (*Sphenodon punctatus*). Lizard-eating birds include the shrikes or butcherbirds, which capture lizards and impale their bodies on thorns or barbed wire fences, roadrunners, egrets, kookaburras, pygmy owls, and a wide range of raptors, both large and small, including the Secretarybird (*Sagittarius serpentarius*), hawks, buzzards, eagles, and falcons. There are many animals that view lizards as juicy snacks, but the biggest threat to wild lizard populations comes from humans (see page 79).

CAMOUFLAGE AND CRYPSIS

Being able to blend into the background is a good way to avoid being eaten, and lizards have perfected this skill in several different ways. Hiding in plain sight is called "crypsis," and the most famous practitioners are the chameleons, which are able to change their color to mimic that of their surroundings. The skin of chameleons contains several layers of pigmented cells known as chromatophores. The outer layer contains the xanthophores, the yellow, orange, or red pigments; next are the iridophores, which are colorless but reflect blue light; and deeper down, the melanophores, which contain black pigment. The melanophores can extend upward between the other layers to darken or obscure the pigmented or reflective layers above, thereby effecting a change of hue or color. Since chromatophores are sensitive to light they will even respond when the chameleon is asleep, darkening it to avoid detection if a light is shone on it.

However, not all chameleons use this elaborate color-changing tactic. The pygmy chameleons (*Rhampholeon*) live in low vegetation close to the ground, and are gray-brown with patterning that resembles dead leaves. If they feel threatened they simply drop to the forest floor and lie still, almost invisible in the surrounding leaf litter. Adopting the subtle shades of their rainforest homes, many lizards utilize crypsis to avoid detection, and this is especially true of lizards living on the trunks of large trees where they feed on ants but are themselves vulnerable to birds. Living in this environment are Madagascan flat-tailed geckos (*Uroplatus*), Asian flying dragons (*Draco*), and South American treerunners (*Plica*). They are all dorsoventrally compressed to hug the trunk, with patterning that resembles bark or lichen. Some flat-tailed geckos take crypsis one stage further, their bodies, legs, tails, and heads being edged with a filamentous fringe which both breaks up their lizard-shaped outline and reduces the gecko's shadow on the trunk underneath.

"Stand still and blend in" is also a tactic adopted by desert lizards. The American horned lizards (*Phrynosoma*) and Australian Hidden Dragon (*Cryptagama aurita*) are squat lizards whose body shape, coloration, and patterning make them perfectly adapted to blend into their stony desert homes, while the long-legged, green and yellow striped Western Australian Superb Dragon (*Diporiphora superba*), the African grass lizards (*Chamaesaura*), and the South American Common Grass Anole (*Anolis auratus*) literally seem to disappear in their grassy habitats.

FIGHT OR FLIGHT

Lizards have evolved a variety of threat postures that may be used to deter potential enemies. The blue-tongued skinks (*Tiliqua*) will gape widely and protrude their broad blue tongues to intimidate anything they consider a threat, while Frilled Lizards (*Chlamydosaurus kingii*) will expand the umbrella-like frill around their heads in an attempt to make themselves look larger and more threatening. Both tactics are effectively bluffs; the lizards have little extra to back them up with—the Frilled Lizard will run, while the blue-tongued skinks will bite, but to little effect. Even so, people in northern Papua New Guinea believe the bite of a blue-tongued skink can kill, so the effect is possibly successful if it can deter interference from the most intelligent creature on the planet.

RIGHT | The Northern Leaf-tailed Gecko (*Saltuarius cornutus*) from Cape York Peninsula, Queensland, is a cryptically patterned rainforest lizard. Even if spotted at a glance it is not possible to tell whether it is facing up or down, and therefore where the head is and which way it will run.

ABOVE | The Frilled Lizard (*Chlamydosaurus kingii*), one of the master bluffers, will expand its umbrella-like frill and gape to make itself appear more threatening. If the bluff fails it will turn and run away bipedally.

RIGHT | This Mediterranean House Gecko (*Hemidactylus turcicus*) has just autotomized its tail voluntarily to escape potential predation. It will eventually regenerate a new tail based on a cartilaginous rod.

When faced with a rival male, or a potential enemy, many of the moderately sized to large lizards (such as chameleons, iguanas, or monitor lizards) will turn sideways, laterally compress their bodies, and extend their legs, all to make themselves look larger and more impressive. They may hiss, gape with their mouths, and, in the case of iguanas and monitors, swish their tails about. The attack, when it comes, could be a bite, or a painful lashing with the whip-like tail. Some of the monitor lizards will even grapple opponents, standing on their hind limbs and wrapping their forelimbs around an enemy as they strive to force the opponent to the floor. Both monitor lizards and iguanas bask and sleep on branches over watercourses, and if a potential predator approaches they simply drop from the branch, plunging into the water, to shelter on the bottom or swim away using their long tails for propulsion. Many arboreal lizards (for example, *Anolis* and *Basiliscus*) sleep on thin branches so that they are alerted by the movements of a predator like a snake coming toward them, whereupon they too drop from their perches.

The beaded lizards (*Heloderma*) possess venom glands and grooved teeth in their lower jaws for the injection of venom. Since they feed on eggs and the nestlings of birds and rodents, they do not really require venom for their prey, so it is most likely it is intended for defense in these slow-moving and otherwise vulnerable desert and dry woodland species.

If all threats of violence fail, then flight is the next option. Frilled Lizards will run away at speed, sprinting to a tree or hole bipedally like the basilisks, which also use flight as a tactic, but across water rather than sand. Flying dragons (*Draco*) will leap from a tree trunk, spread their patagia, and glide as far as 98 ft (30 m) away from the threat, while flying or parachute geckos (formerly *Ptychozoon*, now *Gekko*) also leap into space and glide to safety.

Not all escapes are over ground. Dhab lizards and mastigures (*Uromastyx* and *Saara*) are large, herbivorous, arid-habitat lizards that would be extremely vulnerable to predation if they strayed far from their burrows. If danger threatens they sprint to the burrow and dive inside, and should the enemy decide to follow, it may rush directly into the lizard's exceedingly spiny tail. The arboreal Australian Spiny-tailed Skink (*Egernia depressa*) will wedge itself into a tight crevice and block its retreat with its spiny tail. The Arabian Sandfish (*Scincus mitranus*) has no burrow to retreat to, but it does not need one. Sitting exposed on a desert sand dune, it would appear extremely vulnerable to airborne predation, were it not for its digging abilities. As soon as danger threatens, the Arabian Sandfish dives forward into the loose sand and uses its flattened snout and short limbs to "swim" down deep into the dune.

AUTOTOMY

Autotomy is the casting off of a part of the body by an animal in response to a threat. It is effectively self-amputation, and it would appear to be an extreme last resort strategy to avoid predation. Caudal autotomy, the voluntary shedding of the tail, is a very common tactic employed by a large number of mostly small lizards, and it is reflected in the common and scientific names of a number of species, such as the "glass snake" or Scheltopusik (*Pseudopus apodus*), and the Slow Worm, whose scientific name is *Anguis fragilis*, the "fragile snake." Both these species and many others—in the Gekkota,

Scincomorpha, and Lacertoidea especially, but not exclusively—utilize caudal autotomy as an escape tactic. However, not all lizards can voluntarily break off their own tails, and certainly not those that use their tails for other purposes, including prehensile tails for climbing, such as in the Monkey-tailed Skink (*Corucia zebrata*) and chameleons, or as an offensive (*Iguana* and *Varanus*) or defensive weapon (*Uromastyx* and *Xenosaurus*).

The majority of lizards that practice caudal autotomy are small, and many of them have brightly colored tails, either electric blue (*Emoia caeruleocauda*, *Eumeces*, *Plestiodon*, and *Holaspis*) or bright red (*Morethia*), and often with bright yellow longitudinal stripes directing a predator's attention toward the tail. Vigorous tail movements also distract the predator away from the lizard's more vulnerable body. The lizards that practice this technique exhibit fracture planes across the center of their basal caudal vertebrae. If the tail is grasped, the lizard voluntarily autotomizes it, and the tail will then wriggle furiously for some minutes, distracting the attention of the predator away from the escaping tailless lizard. The blood vessels in the stump will cauterize so the lizard does not bleed out, and over time it will regrow a new tail—this will be formed on a cartilaginous rod extending from the last remaining vertebrae. The cartilage rod has no fracture planes, so should an unlucky lizard need to autotomize its tail a second time the fracture must be in the remaining original tail, closer to the body. The scales of the regenerated tail are also homogeneous in both structure and color.

If a tail is only partially autotomized but remains attached to the lizard and supplied with blood, a second tail may grow from the wound, resulting in a forked tail. These are quite common in nature. Autotomy occurs when the fracture occurs across a vertebra, but if the tail is involuntarily fractured between the vertebrae, this is termed pseudoautotomy and no regeneration takes place.

Possibly more startling than the image of a lizard running away and leaving its tail wriggling on the ground is dermal autotomy. This is a practice adopted by the Indo-Pacific Mutilated Gecko (*Gehyra mutilata*), Madagascan fish-scaled geckos (*Geckolepis*), Asian wonder geckos (*Teratoscincus*), African flat geckos (*Afroedura*), and Seychelles bronze geckos (*Ailuronyx*). The skin of these species is only loosely attached to the underlying body, and it tears and comes off easily if the gecko is grasped by a predator. Wriggling free, the gecko escapes, while the predator tries to deal with a mouthful of sticky skin and scales. Having lost its protective skin, the now pink, naked gecko ought to suffer from infection or water loss, but it will survive to regenerate a new coat of scales.

CHEMICAL DEFENSES

The Australian spiny-tailed geckos (*Strophurus*) and New Caledonian chameleon geckos (*Eurydactylodes*) possess glands in their tails from which they can exude or even squirt fine jets of sticky, foul-smelling fluid with the viscosity of honey. The American horned lizards (*Phrynosoma*) go one stage further. When faced with a potential predator, usually a Coyote (*Canis latrans*) or a feral dog, they are able to squirt blood from the sinuses of their eyes. This tactic is reported to work well with canids but is not used for other predators, such as snakes or birds. It is likely that the lizards are sequestering some of the toxins from their ant diets into the blood to make it especially repellent.

The New Guinea green-blooded skinks (*Prasinohaema*) are a group of five species that do not especially resemble one another, apart from having pale green blood. The color is from a biliverdin-like pigment usually associated with

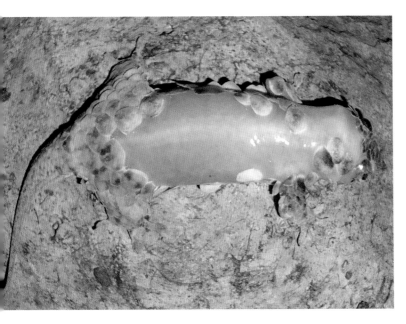

LEFT | The Fish-scaled geckos (*Geckolepis*) are covered in large round scales that resemble those of fish. Any predator attempting to catch one will end up with a mouthful of sticky scales while the gecko lives to grow a new coat of scales.

BELOW | The Regal Horned Lizard (*Phrynosoma solare*) from southwestern USA and northern Mexico, defends itself by squirting jets of blood from sinuses behind the eyes.

the gall bladder, and the obvious purpose for having bitter-tasting green blood would be to deter predation. However, these skinks are arboreal and their predators are birds, which have relatively few taste buds and might not be deterred from eating the lizards if they were swallowing them whole. The reason why these skinks possess this noxious blood chemistry is probably to prevent them from suffering from any of the many lizard malarias, at least three of which occur in New Guinea.

MIMICRY

There are numerous snakes that mimic highly venomous species, such as the mildly venomous false coralsnakes (*Erythrolamprus*), which may mimic the highly venomous true coralsnakes (*Micrurus*), but with few venomous lizards to choose from, any lizards that utilize mimicry must find another model. Juvenile Bushveld Lizards (*Heliobolus lugubris*) are black with yellow spots, but with a tail that is the same color as the sand. They mimic the noxious oogpister "eye-squirter" ground beetles (*Anthia*) in both patterning and movement as a means of avoiding predation. Some geckos, like Carter's Semaphore Gecko (*Pristurus carteri*) or the Western Banded Gecko (*Coleonyx variegatus*) may mimic highly venomous buthid scorpions, because when they feel threatened they turn their tails up over the bodies, while the banded juveniles of the Caatinga Galliwasp (*Diploglossus lessonae*) are believed to resemble millipedes that are toxic to eat because they contain cyanide.

RIGHT | The Armadillo Lizard (*Ouroborus cataphractus*) of South Africa, aims to make itself too spiny and difficult to eat by holdings its tail in its mouth.

BELOW | Carter's Semaphore Gecko (*Pristurus carteri*) makes a good mimic of a highly venomous fat-tailed scorpion (Buthidae, inset).

ARMORED PROTECTION

Most lizards possess armor in the form of their scales, and in some species these overlay protective osteoderms in the skin. The resemblance of their different scalation arrangements to medieval scale armor are no surprise because both permit movement while providing protection. Many lizards also have spiny scales on their bodies, and particularly on their tails, which can be used to block burrows and retreats, preventing a predator from pursuing them. However, one species, the Armadillo Lizard (*Ouroborus cataphractus*), takes protection to the extreme. This is one of the more spinous girdled lizards (Cordylidae), with sharp spiny scales all over its body, head, limbs, and tails, but when it feels threatened it takes its own tail in its mouth and effectively converts itself into a spiny ball, hopefully making itself too large and spiny to consume. The genus name is the name of a dragon from literature and historical iconography, Ouroboros, "the worm that ate its own tail."

CONSERVATION

It might seem a strange thing to say, but extinction is a natural process. Without extinction we surely cannot have the evolution of new taxa; extinction and evolution are two sides of the same coin. There have been five mass extinction events during the last 500 million years. It is believed that the second of these, the Late Devonian extinction (375–360 MYA), resulted in the disappearance of 70 percent of all species on the planet, but it also heralded the evolution, from sarcopterygian fish, of the first tetrapods. The fourth mass extinction event, the Triassic–Jurassic extinction (201 MYA), wiped out most of the large amphibians and therapsid reptiles, thereby leaving terrestrial habitats available for the explosive radiation of the dinosaurs that would occur during the Jurassic. The Cretaceous–Paleogene extinction (66 MYA) brought the reign of the dinosaurs to an end, and allowed some of the surviving therapsids to evolve into the mammals of today, and essentially brought about our own existence.

Naturally occurring extinction is like fire—it clears the chaff and allows the new crop to germinate—but that does not mean that the human-mediated acceleration of species extinction is a good thing for the planet. We are talking about extinction on a geological timescale, not over a few human lifetimes; that is, several million years as

LEFT | The Rodrigues Giant Gecko (*Phelsuma gigas*), one of the lizard species lost for ever. It was likely an important pollinator of island plants which may also potentially have followed it into extinction.

opposed to a few hundred years or even decades. The human-mediated extinction of numerous species that we are experiencing, from geckos to tigers, needs to be controlled, and stopped if possible. We may indeed be heading toward the predicted sixth mass extinction, but it makes no sense to run toward it, and the removal of a single species from nature can have a deleterious effect on the ecology of its former ecosystem and thereby exacerbate the whole process.

For example, the day geckos (*Phelsuma*) are recognized as "keystone species" because of their importance to Indian Ocean island ecosystems. On the islands of Mauritius and Réunion, the endangered and endemic flowering shrubs of genus *Trochetia* produce red or yellow nectar to attract geckos and passerine birds called white-eyes (*Zosterops*). Rather than insects, it is the day geckos and white-eyes that pollinate these flowers, and their extinction could lead to the extinction of the plants, and then all the species that rely upon those plants. Rodrigues Island, to the east of Mauritius, was also once home to two *Phelsuma* species. The nocturnal Rodrigues Giant Gecko (*P. gigas*) went extinct in 1841 and the smaller Rodrigues Day Gecko (*P. edwardnewtoni*) followed in 1917. Despite the introduction of the Mauritius Blue-tailed Day Gecko (*P. cepediana*) to the island, it is impossible to say how the loss of the two native geckos may have affected the native flora, and ultimately the island's entire ecology.

Lizards have many enemies, some natural, and others introduced and therefore unnatural. However, the greatest enemy, whether directly or indirectly, must be *Homo sapiens*, our own species. Certainly, island lizard species may have gone extinct in the past due to hurricanes in the Caribbean, cyclones in the Indian Ocean, or typhoons in the western Pacific. In centuries gone by these climatic events did not involve humans, but today our actions even affect the planet's weather patterns, so we can no longer divorce ourselves from responsibility for the freak weather that causes all too frequent and widespread devastation, not only for our own species but also for the natural world. The Australian mammalogist, climate scientist, and author Tim Flannery referred to our capacity to use up all the resources that would otherwise sustain the planet for generations to come as "future eating."

FIRE

The devastating 2019–2020 Australian and Californian bushfire season was one of the worst on record. Yet fire is a component of nature, and many animals and plants in areas affected by fire have, since prehistory, evolved to survive it. Some seeds have germination inhibitors in their outer coats so that they will not germinate until a fire has passed by, scorched off the outer coat, and burned all the dead vegetation, so when the seedlings germinate they will have the sunlight they require for photosynthesis. However, the fires that Australia and California experienced burned longer, hotter, and deeper than those that benefit nature; even trees with bark adapted to withstand the passing of quick bushfires did not survive. Look at any field guide to Australian lizards, snakes, invertebrates, frogs, birds, or mammals and see how many species have small, localized ranges that lie entirely within the recent fire zones, and then wonder how many of those species may have gone extinct during the 2019–2020 bushfires.

FERAL AND INVASIVE SPECIES

Oceanic islands are effectively isolated microcosms of nature that are often not affected by changes occurring on the mainland. However, when an invasive species or other threat reaches an oceanic island the effects can be catastrophic. The ability of our species to build seaworthy boats has been the agency for bringing such devastation to isolated islands all around the world. Humans arrive and either hunt the island species, or change their habitat in such a way that they can no longer survive and, being on an island, have nowhere else to go. Also, humans rarely arrive alone but bring with them cats, goats, or pigs, and unsuspectingly introduce rats, the Island Wolfsnake (*Lycodon capucinus*), and other species that can have a very negative effect on the island's naive native wildlife. There are many accounts of such invasions bringing about island extinctions, the best documented being those that affected birds, such as the late 1940s arrival of the Brown Treesnake (*Boiga irregularis*) on Guam, where it rapidly ate several bird species into extinction, or the 1894 arrival of a new lighthouse keeper, and his pet cat, which had kittens, on Stephens Island off the coast of New Zealand. In just over a year the flightless Lyall's Wren (*Traversia lyalli*) was extinct.

One of the earliest possible consequences of humans colonizing a new land occurred when the first humans reached Australia around 65,000 years ago. At that time the dominant terrestrial reptile was probably the Megalania (*Varanus priscus*), a giant monitor lizard. Although authorities disagree about its maximum size, with lengths estimated from 18 to 26 ft (5.5 to 7.9 m), Megalania was much larger than our largest extant lizard, the Komodo Dragon (*V. komodoensis*). For all its potential fearsomeness, Megalania went extinct around 48,000 years ago, and this is thought to have been caused either through direct persecution, hunting, or competition with humans.

The great exploring navies of the seventeenth, eighteenth, and nineteenth centuries possessed vessels that could circumnavigate the globe and remain at sea for years. These ships would stop at islands to resupply with fresh water and meat. Island species without predators were seemingly

LEFT | The Round Island Skink (*Leiolopisma telfairii*) only survives on Round Island in the Indian Ocean, because the island is now a protected reserve.

RIGHT | The Round Island Day Gecko (*Phelsuma guentheri*) is another lizard saved from the brink of extinction by the listing of its island home as a biological reserve.

not afraid of humans, and many of the birds were flightless. Species like the Dodo (*Raphus cucullatus*) on Mauritius, the Solitaire (*Pezophaps solitaria*) on Rodrigues Island, and the giant tortoises of the Seychelles (*Aldabrachelys*) and Mascarene Islands (*Cylindraspis*) were easy game. Tortoises were valued as they would survive for long periods onboard ship to provide fresh meat on the following voyage. Sailors would also have hunted large lizards for food, such as the endemic island iguanas (*Cyclura*) of the Caribbean. And when the ships left the island the devastation they had wrought would not end because they left rats, cats, and goats behind.

Mauritius had such a serious rat problem that the Javan Mongoose (*Herpestes javanicus*) was introduced to hunt them. Unfortunately, the mongooses only helped to speed elements of the native Mauritian fauna toward their extinctions. Today a microcosm of the original Mauritian herpetofauna survives on the tiny Round Island 14 miles (22.5 km) to the north, but this too was almost lost before it was declared a nature reserve and given protection in the 1970s. Round Island is the last native home to three rare reptiles, the Round Island Day Gecko (*Phelsuma guentheri*), Round Island Skink (*Leiolopisma telfairii*), and Round Island Keel-scaled Boa (*Casarea dussumieri*), but the conservation efforts came too late for another snake, the Round Island Burrowing Boa (*Bolyeria multocarinata*), which has not been seen since 1975.

A similar situation occurred on New Zealand, which was first colonized by the Maori in the fourteenth century, with Europeans arriving 300 years later. Until that time New Zealand was truly isolated and its flora and fauna were pristine, but with the arrival of humans came rats, and with the Europeans came cats, dogs, ferrets, stoats, and foxes. Soon vulnerable species like the world's only flightless parrot, the Kakapo (*Strigops habroptilus*), and the last rhynchocephalian, the Tuatara (*Sphenodon punctatus*, page 88), were disappearing across mainland New Zealand; today they only survive in the wild on closely monitored, predator-free islands off the coast. The Kawekaweau (*Hoplodactylus delcourti*) was the world's largest gecko, until the nineteenth century. The only known specimen, a stuffed museum exhibit in

Marseille, has a SVL of 14½ in (370 mm), much larger than the 11 in (280 mm) of Leach's Giant Gecko (*Rhacodactylus leachianus*) from New Caledonia, the current record holder. The last Kawekaweau encountered was reportedly killed by a Maori chief who found it under some bark in 1870, but it seems likely the writing was already on the wall for this large yet defenseless lizard due to the deluge of invasive species arriving from England.

LIZARD EXTINCTIONS

At least 24 species and subspecies of lizards have been documented as having gone extinct, or probably extinct, during the last 420 years, although the true figure is probably much higher, including species we never knew existed and to which scientists had not even applied names. If these known extinctions are plotted on a world map, a very clear pattern emerges, with most of the extinctions centered in two areas: the Caribbean islands (with nine species) and the western Indian Ocean islands (six species). However, there are also extinct species recorded for the Cape Verde Islands, Mediterranean islands, Christmas Island, Tonga, and New Zealand (the aforementioned Kawekaweau) (see table on pages 84-5). These are mostly island extinctions; few are continental, the list including just one from Uruguay and two from South Africa, but that is not to say continental species have not gone extinct—it is just harder to confirm unless the species had a very small and specific range. Another interesting aspect is that many of the extinct island endemics were giants (gigantism and dwarfism are typical characteristics of animals on islands), including not just the largest known gecko, but also the largest known skink, the Cape Verde Giant Skink (*Chioninia coctei*).

Of particular relevance to the present day is the situation on Christmas Island, a small Australian External Territory in the Indian Ocean, south of Java. This small island was home to five endemic reptiles: the Christmas Island Snake-eyed Skink (*Cryptoblepharus egeriae*), Christmas Island Emo Skink (*Emoia nativitatis*), Christmas Island Bent-toed Gecko (*Cyrtodactylus*

ABOVE | The Cape Verde Giant Skink (*Chioninia coctei*) went extinct due to hunting and a prolonged drought.

ABOVE | Christmas Island issued a set of stamps to recognize the island's endemic, endangered, or extinct herpetofauna.

sadlieri), Christmas Island Chained Gecko (*Lepidodactylus listeri*), and Christmas Island Blindsnake (*Ramphotyphlops exocoeti*). Although phosphate mining between 1899 and 1988 may have negatively affected some populations, and they may also have been impacted by the construction of an immigration detention center in 2006, the main threat to the survival of these small, endemic insular reptiles came from the accidental introduction of the voracious Yellow Crazy Ants (*Anoplolepis gracilipes*) that will consume anything they can capture. This relatively large ant is so-named because of its erratic movements while foraging. Originating from Southeast Asia, it has invaded Australia and islands across the Indian and Pacific Oceans, forming supercolonies that cause ecological catastrophes wherever they occur.

The situation would have been further exacerbated on Christmas Island by the introduction of the lizard-eating Island Wolfsnake around 1987, and two perianthropic geckos, the Asian House Gecko (*Hemidactylus frenatus*) and Mutilated Gecko (*Gehyra mutilata*), which represented serious competition for the smaller native geckos. Invasive cats and rats are also present on the island, as is the giant centipede *Scolopendra subspinipes*, which will also predate small vertebrates.

Most (63 percent) of Christmas Island is now designated as a national park, but the prospects for the island's herpetofauna are worrying. The Christmas Island Emo Skink is now listed as Extinct by the IUCN (International Union for Conservation of Nature), while the Christmas Island Snake-eyed Skink and Christmas Island Chained Gecko are listed as Extinct in the Wild, though there are successful captive breeding programs ongoing for both species. The Christmas Island Bent-toed Gecko and Christmas Island Blindsnake are listed as Endangered. Christmas Island was, and is, also home to other endemic vertebrates, including two rats, two bats, a shrew, and several endemic birds, which are today listed as Near Threatened, Vulnerable, Endangered, Critically Endangered, or Extinct.

CONSERVATION PROJECTS

Around the world, intensive conservation programs have been established to help save endangered island lizard species. One that has proven to be very successful is the multi-agency "Blue Iguana Recovery Program" in the Cayman Islands. The Grand Cayman Blue Iguana (*Cyclura lewisi*) inhabits Grand Cayman in the Greater Antilles, to the west of Jamaica. The genus *Cyclura* contained 11 species, but the Navassa Rhinoceros Iguana (*C. onchiopsis*) went extinct in the late nineteenth century (see table on pages 84-5). Of the remaining ten species, two are Vulnerable, five Endangered, and three Critically Endangered. The Grand Cayman Blue Iguana is listed as Endangered but increasing in numbers because of the recovery program, but prior to the program's initiation the wild population was down to 50 individuals, and it continued to fall to between five and 15 individuals by 2003. The wild iguanas were threatened with extinction due to the presence of rats, which ate their eggs, feral cats and mongooses, which attacked the juveniles, dogs, which hunted adults, and goats and Cactus Moths (*Cactoblastis cactorum*), which ate the food of the herbivorous adult iguanas. Habitat destruction or alteration, increased road traffic, trapping, and competition from introduced Green Iguanas (*Iguana iguana*) also threatened the wild population.

The captive breeding program, which began in 1990 with 30 adult Grand Cayman Blue Iguanas, has been successful, and the aim is to reintroduce 1,000 iguanas back into the wild. However, these reintroductions are only going to be successful in the long term if the invasive predators and competitors have been eradicated, or at the very least controlled. The National Trust for the Cayman Islands has now turned its attention and expertise toward the Critically Endangered Lesser Caymans Iguana (*Cyclura nubila caymanensis*), which is down to 150 individuals on Cayman Brac and 1,500 individuals on Little Cayman.

BELOW | The aptly named Grand Cayman Blue Iguana (*Cyclura lewisi*) has been the recipient of one of the most successful island species recovery plans for a lizard. Now the methodology can be transferred to other endangered island species.

LIZARD SPECIES THAT HAVE GONE EXTINCT OVER THE LAST 500 YEARS

1
Jamaican Giant Galliwasp
(*Celestus occiduus*) 1872

2
Navassa Curlytail
(*Leiocephalus eremitus*) 1860s
Navassa Ground Iguana
(*Cyclura onchiopsis*) 19th century

3
Martinique Curlytail
(*Leiocephalus herminieri*) 1840
St Lucia Skink (*Alinea luciae*) 1879
Martinique Giant Ground Lizard
(*Pholidoscelis major*) 19th century

4
Marie-Galante Ground Lizard
(*Pholidoscelis turukaeraensis*)
17th century
Leeward Islands Curlytail
(*Leiocephalus cuneus*) 17th century
Redonda Skink (*Copeoglossum redondae*) 1873
Guadeloupian Giant Ground Lizard
(*Pholidoscelis major*) 19th century
Guadeloupian Ground Lizard
(*Pholidoscelis cineraceus*) 1928

5
Cabo Polonio Whiptail
(*Contomastix charrua*) 1977

6
Cape Verde Giant Skink
(*Chioninia coctei*) 1898

11
Réunion Giant Skink
(*Leiolopisma cecilae*) 1670
Réunion Slender-toed Gecko
(*Nactus soniae*) 17th century
Réunion Skink (*Gongylomorphus bojerii borbonicus*) 1840

12
Mauritius Giant Skink
(*Leiolopisma mauritana*) 1603

13
Rodrigues Giant Gecko
(*Phelsuma gigas*) 1842
Rodrigues Day Gecko
(*Phelsuma edwardnewtoni*) 1917

7
Ratas Island lizard
(*Podarcis lilfordi rodriguezi*) 1950

8
San Stephano Island lizard
(*Podarcis siculus sanctistephani*) 1965

9
Eastwood's long-tailed seps
(*Tetradactylus eastwoodae*) 1928

10
Günther's dwarf burrowing skink
(*Scelotes guentheri*) 1887

14
Christmas Island Emo Skink
(*Emoia nativitatis*) 2010

15
Tongan Giant Skink
(*Tachygyia microlepis*) 19th century

16
Kawekaweau (*Hoplodactylus delcourti*)
18th century

THE LIZARD INFRAORDERS

Given the amount of time lizards have been on Earth, and the wide array of habitats and lifestyles they have evolved to occupy, it is not surprising that they demonstrate a high degree of diversification, with six distinct lineages, which may be treated as infraorders. The Dibamia, containing just the Dibamidae, a small family of near-blind and almost limbless burrowing lizards, is considered by many to be the sister-clade to all other squamate reptiles (lizards, worm-lizards, and snakes). The remaining five infraorders are listed opposite: Gekkota (geckos and their allies); Scincomorpha (skinks and their allies); Lacertoidea (typical lizards and the amphisbaenians or worm-lizards); Iguania (agamas, chameleons, and iguanians), and Anguimorpha (slow worms to monitor lizards), with the snakes believed most closely related to this last infraorder. Even though there may be disagreement over which lizard families are most basal, most authors continue to group the families together in these infraorders.

BELOW | A juvenile Veiled Chameleon (*Chamaeleo calyptratus*), with its turret-eyes; zygodactylous feet; prehensile tail; elongate, rapid-fire, adhesive tongue, and the capacity for sudden color-changing, epitomizes just one of the amazing extremes of lizard evolution.

DIBAMIA — DIBAMIDAE (Blind-lizards)

GEKKOTA
- CARPHODACTYLIDAE (Southern padless geckos)
- DIPLODACTYLIDAE (Austral geckos)
- PYGOPODIDAE (Flap-footed lizards)
- GEKKONIDAE (Cosmopolitan geckos)
- EUBLEPHARIDAE (Eyelid geckos)
- PHYLLODACTYLIDAE (Leaf-toed geckos)
- SPHAERODACTYLIDAE (Dwarf & least geckos)

SCINCOMORPHA
- CORDYLIDAE (Girdled lizards & Flat lizards)
- GERRHOSAURIDAE (Plated lizards)
- XANTUSIIDAE (Night lizards)
- SCINCIDAE (Skinks)

LACERTOIDEA
- **TEIIOIDEA**
  - TEIIDAE (Teiids & Tegus)
  - ALOPOGLOSSIDAE (Shade teiids)
  - GYMNOPHTHALMIDAE (Microteiids)
  - LACERTIDAE (Old World lizards)
- **AMPHISBAENIA**
  - AMPHISBAENIDAE (Worm-lizards)
  - BLANIDAE (Mediterranean worm-lizards)
  - BIPEDIDAE (Ajolates)
  - CADEIDAE (Cuban worm-lizards)
  - RHINEURIDAE (Florida worm-lizard)
  - TROGONOPHIDAE (Afro-Arabian worm-lizards)

IGUANIA
- **ACRODONTA**
  - AGAMIDAE (Agamas & Dragons)
  - CHAMAELEONIDAE (Chameleons)
- **PLEURODONTA**
  - IGUANIDAE (Iguanas & Chuckwallas)
  - LEIOCEPHALIDAE (Curlytails)
  - HOPLOCERCIDAE (Woodlizards, Manticores & Weapontail)
  - CROTAPHYTIDAE (Collared & Leopard lizards)
  - CORYTOPHANIDAE (Basilisks)
  - LEIOSAURIDAE (South American Tree & Ground lizards)
  - LIOLAEMIDAE (South American Swifts & Tree lizards)
  - POLYCHROTIDAE (Bush anoles)
  - DACTYLOIDAE (True anoles)
  - PHRYNOSOMATIDAE (Spiny & Horned lizards)
  - TROPIDURIDAE (Lava lizards, Whorltails & Treerunners)
  - OPLURIDAE (Malagasy swifts & Iguanines)

ANGUIMORPHA
- **DIPLOGLOSSA**
  - ANGUIDAE (Slow worms, Alligator lizards & Glass lizards)
  - XENOSAURIDAE (Knob-scaled lizards)
  - DIPLOGLOSSIDAE (Galliwasps)
  - SHINISAURIDAE (Chinese crocodile lizard)
- **PLATYNOTA**
  - HELODERMATIDAE (Gila monster & Beaded lizards)
  - LANTHANOTIDAE (Borneo earless monitor lizard)
  - VARANIDAE (Monitor lizard)

SPHENODONTIDAE
TUATARA

LEFT | Although the Tuatara may resemble a gray-brown lizard, it is anything but; it is the sister taxon of all squamates—the snakes, lizards, and worm-lizards.

The Tuatara (*Sphenodon punctatus*) of New Zealand is the last surviving member of the order Rhynchocephalia (beak-headed reptiles), which in the Triassic occupied Europe, Asia, Africa, and North and South America, but became extinct across most of their range during the Jurassic or Cretaceous. The Rhynchocephalia is the sister taxon to the Squamata (snakes, lizards, and worm-lizards) and together they form the Lepidosauria (scaled reptiles). Today, even in remote New Zealand, Tuataras are under threat from invasive cats and rats and only survive on a few islands, although mainland reintroductions are being conducted. "Tuatara" is a Maori word meaning "spiny-backed."

Tuataras have a distinctive pineal eye, lack external ear openings, and exhibit a transverse cloacal opening like snakes and lizards, but males lack a copulatory organ. Instead they possess two shallow pockets in their cloacas, which are everted during mating and may just enter the female to aid sperm transfer.

Tuataras have slow metabolisms, are not sexually mature until 13–17 years old, but may live to be 60 or 100. They inhabit coastal scrub, forest, or grassland and often dwell in seabird burrows. Nocturnal in habit, they feed on insects, weta, centipedes, small lizards, and seabird eggs and chicks. Females only produce a clutch of eggs every four years and these take 16 months to hatch.

DISTRIBUTION
New Zealand

GENUS
Sphenodon

HABITATS
Coastal scrub, grassland, and forest

SIZE
SVL 11 in (280 mm) Tuatara (*Sphenodon punctatus*)

ACTIVITY
Terrestrial and semi-fossorial; nocturnal

REPRODUCTION
Oviparous, producing a clutch of 1–18 eggs every four years

DIET
Arthropods, including beetles, weta, and centipedes; also snails, geckos, skinks, birds' eggs, and chicks

DIBAMIDAE
BLIND-LIZARDS

There are several possible phylogenies for the Squamata but in the one adopted here the Dibamidae is the sister taxa to all other lizards, worm-lizards, and snakes. The family contains two genera and 25 species, and it exhibits an unusual distribution. Most of the species are in the genus *Dibamus*, inhabiting southern China and Southeast Asia, often with very localized distribution, but a single species, the Mexican Blind-lizard (*Anelytropsis papillosus*), occurs in northeastern Mexico.

Dibamids are elongate serpentine lizards with small, close-fitting, glossy scales and superficially they resemble blindsnakes (Typhlopidae) or threadsnakes (Leptotyphlopidae). They completely lack forelimbs, but closer examination will reveal a pair of vestigial hind limbs, reduced to small, scaly flaps. External ear openings are absent and the eyes are vestigial. They can also autotomize (break off) their tails in defense, a defensive adaptation not seen in blindsnakes or threadsnakes.

These are secretive fossorial reptiles that are rarely encountered on the surface, but may be found under logs. While *Dibamus* occurs in tropical rainforest habitats, *Anelytropsis* has adapted to live in dry upland forest and scrub. *Dibamus* is oviparous, lays single eggs, and feeds on insects and possibly earthworms. *Anelytropsis* is largely unstudied.

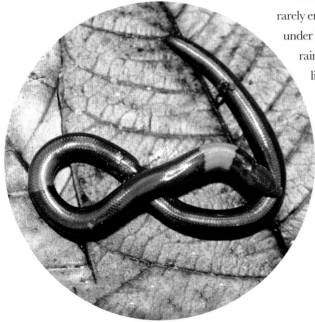

LEFT | The Nicobar worm-lizard (*Dibamus nicobaricum*) superficially resembles a blindsnake but its head is distinctly that of a lizard.

DISTRIBUTION
Southeast Asia and northeastern Mexico

GENERA
Anelytropsis and *Dibamus*

HABITATS
Tropical rainforest, dry tropical forest, and pine–oak forest

SIZE
SVL 3⅓ in (86 mm) Greers' Blind-lizard (*Dibamus greeri*) to 8 in (203 mm) Seram Blind-lizard (*D. seramensis*)

ACTIVITY
Semi-fossorial to fossorial or inside rotten tree trunks; probably nocturnal

REPRODUCTION
Oviparous, laying single eggs; some species may multi-clutch (*Dibamus*); *Anelytropsis* unknown

DIET
Arthropods, including insect larvae, and also earthworms (*Dibamus*); *Anelytropsis* unknown.

LEFT The Tokay Gecko (*Gekko gecko*) is the largest member of the largest gekkotan family Gekkonidae.

INFRAORDER GEKKOTA

The infraorder Gekkota contains seven families of geckos. Three are centered on Australia, from where the Diplodactylidae (Austral Geckos) range to New Caledonia and New Zealand, while the Pygopodidae (scaly-feet) also occur in New Guinea. The Carphodactylidae (southern padless geckos) are endemic to Australia.

The other four families are more widely distributed. The Eublepharidae (eyelid geckos) inhabit Asia, East and West Africa, and North America. The Sphaerodactylidae (dwarf or least geckos) and Phyllodactylidae (leaf-toed geckos) inhabit the Americas, Europe, and Western Asia. Phyllodactylids also occur on Madeira, the Canary Islands, Cape Verde, and Socotra.

The largest gecko family, the Gekkonidae (cosmopolitan geckos), is distributed worldwide, except Antarctica, and on islands ranging in size from New Guinea to tiny Pacific Ocean atolls. It contains more than 66 percent of all geckos. Most of the commensal species that live alongside humans are found in this family, as are the best colonists.

All limbed gekkotan families contain both dilated, leaf-toed species, and slender, bent-toed species. Most are oviparous, but many diplodactylids are viviparous. Geckos usually lay a pair of eggs, but very small species lay single eggs.

CARPHODACTYLIDAE
SOUTHERN PADLESS GECKOS

All members of the endemic Australian gecko family Carphodactylidae have curiously shaped, often spinous, tails and either elongate or broad heads. They also lack the dilated terminal toe pads seen in many typical geckos. The knob-tailed geckos (*Nephrurus*) possess short tails, often swollen at the base, which terminate in a small rounded knob, but other genera have very different tail shapes.

BELOW | The Northern Knob-tailed Gecko (*Nephrurus sheai*) has a short, bulbous tail that terminates in a small, round knob. It is a nocturnal and terrestrial inhabitant of relatively arid habitats.

DISTRIBUTION
Australia

GENERA
Carphodactylus, Nephrurus, Orraya, Phyllurus, Saltuarius, Underwoodisaurus, and *Uvidicolus*

HABITATS
Rainforest to rocky outcrops and desert

SIZE
SVL 2¾ in (70 mm) Granite Belt Thick-tailed Gecko (*Uvidicolus sphyrurus*) to 6⅓ in (160 mm) Northern Leaf-tailed Gecko (*Saltuarius cornutus*)

ACTIVITY
Arboreal, terrestrial, or saxicolous; all species are nocturnal

ABOVE | The McIlwraith Leaf-tailed Gecko (*Orraya occulta*) inhabits large boulders along rainforest creeks, and with its cryptic patterning and shape it easily escapes notice.

Some leaf-tailed geckos (*Saltuarius*) have broad, flattened tails while the broad-tailed geckos (*Phyllurus*) have bulbous tails that taper to a fine point.

Many carphodactylids are inhabitants of deserts and other arid habitats such as spinifex grasslands or vegetated sand dunes, but the secretive Australian Chameleon Gecko (*Carphodactylus laevis*) only occurs in the Queensland rainforests, where it forages for invertebrates in the leaf litter and low vegetation at night. Many other species also exhibit extremely localized ranges, such as the McIlwraith Leaf-tailed Gecko (*Orraya occulta*), which is confined to one Queensland mountain range, and the Border Thick-tailed gecko (*Uvidicolus sphyrurus*), from northeastern New South Wales and southern Queensland.

Whether desert or rainforest inhabitants, the pastel colors and unusual tail and body shapes of these geckos enable them to blend in perfectly with their habitats and, if motionless, escape detection. The knob-tailed geckos and thick-tailed geckos (*Underwoodisaurus*) adopt a defensive tactic that involves extending their legs, so that they stand high off the ground, before lunging open-mouthed at a potential threat, and uttering a loud bark. This has led to some species, such as *U. seorsus*, being called barking geckos.

REPRODUCTION
All species are oviparous, laying a pair of parchment-shelled eggs

DIET
Arthropods, primarily insects and spiders; some desert species take scorpions

DIPLODACTYLIDAE
AUSTRALIAN GECKOS

The Austral geckos comprise 154 species that also inhabit New Zealand and New Caledonia, but Australia is home to the lion's share—98 species in ten genera. Most species have toes that are dilated into toe pads for climbing, but some species exhibit slender, non-dilated toes, such as the terrestrial Beaded Gecko (*Lucasium damaeum*). Australia's smallest geckos, the clawless geckos (*Crenadactylus*), are terrestrial spinifex-grassland lizards that completely lack claws and, like the Northern Phasmid Gecko (*Strophurus taeniatus*), have elongate bodies and striped yellowish tails to help them blend into their surroundings.

While many species have smooth, velvety skin and slender tails, like the Gracile Velvet Gecko (*Oedura gracilis*) and Clouded Velvet Gecko (*Amalosia jacovae*), or bulbous tails, like many of the sedentary stone geckos (*Diplodactylus*), a few species

LEFT | At up to 4¾ in (120 mm) SVL the rainforest-dwelling Giant Tree Gecko (*Pseudothecadactylus australis*) is the largest Australian diplodactylid.

RIGHT | The Gulf Marbled Velvet Gecko (*Oedura bella*), which was described in 2016 from Queensland and Northern Territory, has soft velvety skin and a bulbous tail.

DISTRIBUTION
Australia, New Caledonia, and New Zealand

GENERA
Amalosia, Bavayia, Correlophus, Crenadactylus, Dactylocnemis, Dierogekko, Diplodactylus, Eurydactylodes, Hesperoedura, Hoplodactylus, Lucasium, Mniarogekko, Mokopirirakau, Naultinus, Nebulifera, Oedodera, Oedura, Paniegekko, Pseudothecadactylus, Rhacodactylus, Rhynchoedura, Strophurus, Toropuku, Tukutuku, and Woodworthia.

HABITATS
Desert, rocky outcrops, rainforest, woodland, heathland, grassland, and islands

have raised tubercles on the back, and tails fringed with spines, such as the Western Shield Spiny-tailed Gecko (*Strophurus wellingtonae*).

Most Australian diplodactylids are terrestrial inhabitants of sandy or stony desert, and many, including the beaked geckos (*Rhynchoedura*), shelter in burrows made by insects or spiders. Other species inhabit dry sclerophyll woodland, such as the Robust Velvet Gecko (*Nebulifera robusta*), and a few species inhabit large trees or rocky caves, such as the Giant Tree Gecko (*Pseudothecadactylus australis*) and Western Giant Cave Gecko (*P. cavaticus*). These two species are the largest of Australian diplodactylids (over 4 in/100 mm SVL), and they possess broad toe pads and an adhesive underside to the tip of their prehensile tails to aid climbing. All Australian diplodactylids are oviparous, laying a pair of leathery-shelled eggs.

SIZE
SVL 1 in (28 mm) Pilbara Clawless Gecko (*Crenadactylus pilbarensis*) to 11 in (280 mm) Leach's Giant Gecko (*Rhacodactylus leachianus*). The Kawekaweau (*Hoplodactylus delcourti*) is a presumed extinct gecko of 14½ in (370 mm) SVL thought to have been collected in New Zealand

ACTIVITY
Arboreal, terrestrial, or saxicolous; all species are nocturnal

REPRODUCTION
All Australian and the majority of New Caledonian species are oviparous, laying a pair of parchment-shelled eggs, except New Caledonian chameleon geckos (*Eurydactylodes*), which lay two hard-shelled eggs. All New Zealand geckos and two New Caledonian *Rhacodactylus* are viviparous, producing 1–2 neonates.

DIET
Arthropods; primarily insects and spiders, but larger species, such as *R. leachianus*, take smaller lizards or birds, and the Stephen's Sticky-toed Gecko (*Toropuku stephensi*) also feeds on nectar, flowers, or fruit.

DIPLODACTYLIDAE
NEW ZEALAND & NEW CALEDONIAN GECKOS

The diplodactylids are the dominant geckos in New Zealand, with at least 20 species in seven genera, while New Caledonia is home to 36 species in eight genera. All New Zealand species are viviparous, an advantage for reptiles living in cool habitats, whereas all the New Caledonian species are oviparous, except the Rough-headed Giant Gecko (*Rhacodactylus trachycephalus*) and the Rough-snouted Giant Gecko (*R. trachyrhynchus*).

The largest New Zealand genus is *Naultinus*, with eight species of mostly brilliant green tree geckos; for example, the Elegant Gecko (*N. elegans*), which also occurs in a black-speckled xanthic (yellow) phase. While these tree geckos have the most slender toes, the New Zealand geckos with the most dilated toes are the sticky-toed geckos (*Woodworthia*). The largest gecko in New Zealand is Duvaucel's Gecko (*Hoplodactylus duvauceli*), which achieves a SVL of 6½ in (165 mm), but is mostly confined to offshore islands. The only larger gecko would have been the presumed extinct Kawekaweau (*H. delcourti*), at 14½ in (370 mm). New Caledonia's largest gecko is Leach's Giant Gecko (*Rhacodactylus leachianus*), at 11 in (280 mm) SVL, it is also the world's largest living gecko.

New Caledonia is home to a few unusual geckos, such as the Crested Gecko (*Correlophus ciliatus*), which has an angular head and a spiny crest over the eyes, around the head, and down the back, and the Large-scaled Chameleon Gecko

LEFT | The Jewelled Gecko (*Naultinus gemmeus*) from Otago, South Island, has a brilliant green and pale-striped pattern, but other populations are mottled green, gray, or brown.

BELOW | Leach's Giant Gecko (*Rhacodactylus leachianus*), from New Caledonia, is the largest living gecko, despite its extremely short tail.

(*Eurydactylodes symmetricus*), with its head covered in curious large, symmetrical scales. Many of the New Caledonian geckos also possess adhesive tail tips to aid in climbing.

There are potentially many more New Zealand and New Caledonian species to be described from the small New Caledonian genera *Bavayia* and *Dierogekko*, and the Pacific Gecko of New Zealand (*Dactylocnemis pacificus*), which may be a species complex.

PYGOPODIDAE—LIALISINAE & PYGOPODINAE
FLAP-FOOTED LIZARDS

The pygopodids are not strictly legless; they retain vestigial hind limbs as scaly flaps, but do completely lack forelimbs. They are divided into two subfamilies: three genera and 27 species in the Pygopodinae (scaly-foots and flick-leapers), which have large scaly flaps, and four genera and 19 species in the Lialisinae (snake-lizards and worm-lizards), which have small scaly flaps. All but one species of *Aprasia*, the Eared Worm-lizard (*A. aurita*), also lack any external ear openings.

Most pygopodids are endemic to continental Australia, except the genus *Lialis*, which comprises two species—Burton's Snake-lizard (*L. burtonis*), the most widespread Australian reptile, which also inhabits southern New Guinea, and Jicar's Snake-lizard (*L. jicari*), which is endemic to New Guinea.

LIALISINAE

DISTRIBUTION (RED AND GREEN ON MAP)
Australia and New Guinea

GENERA
Aprasia, *Lialis*, *Ophidiocephalus*, and *Pletholax*

HABITATS
Desert, grassland, mallee, woodland, and sclerophyll forests

SIZE
SVL 3½ in (90 mm) Keeled Legless Lizard (*Pletholax gracilis*) or East Wallabi Island Worm-lizard (*Aprasia clairae*) to 12⅓ in (314 mm) Jicar's Snake-lizard (*Lialis jicari*)

ACTIVITY
Terrestrial, fossorial, or semi-fossorial; diurnal or nocturnal

REPRODUCTION
Oviparous, producing pairs of eggs, although Burton's Snake-lizard (*Lialis burtonis*) may produce 1–3 eggs

LEFT | Burton's Snake-lizard (*Lialis burtonis*) is the most widely distributed reptile in the Australo-Papuan region. Its long jaws are highly flexible to enable it to capture, grasp, kill, and consume skinks.

ABOVE | The Common Scaly-foot (*Pygopus lepidopodus*) resembles a snake and even acts like a snake when it feels threatened.

Pygopodids are often mistaken for snakes, and this is put to good effect by the scaly-foots (*Pygopus*) and the Brigalow Scaly-foot (*Paradelma orientalis*), which elevate the anterior portion of their bodies and flick their tongues when threatened, possibly mimicking a venomous snake. Most species move with serpentine motion, but the flick-leapers (*Delma*) are so-called because they can leap across the ground by flicking the body forward.

Pygopodids inhabit arid habitats, but they are most common in tussock or spinifex grasslands, and although they are terrestrial, the Slender Slider (*Pletholax gracilis*) has been observed climbing trees. Many species are extremely localized in their distribution and vulnerable to habitat alteration, and the Bronzeback (*Ophidiocephalus taeniatus*), which inhabits desert scrublands, is considered endangered in parts of its range.

Most pygopodids prey on arthropods, from ant larvae, eggs, and pupae taken by the tiny Australian worm-lizards (*Aprasia*) to spiders and scorpions taken by the robustly built scaly-foots (*Pygopus*). The snake-lizards (*Lialis*) prey on lizards, especially skinks, which are captured and held in their vise-like jaws.

DIET
Primarily arthropods, *Aprasia* feeding on ant eggs, pupae, and larvae, but *Lialis* preys on skinks

PYGOPODINAE

DISTRIBUTION (RED AND BLUE ON MAP)
Australia

GENERA
Delma, *Paradelma*, and *Pygopus*

HABITATS
Desert, grassland, mallee, woodland, and rocky outcrops

SIZE
SVL 2½ in (63 mm) Adorned Delma (*Delma torquata*) to 10¾ in (274 mm) Common Scaly-foot (*Pygopus lepidopodus*)

ACTIVITY
Terrestrial, fossorial, or semi-fossorial, though *Paradelma* may be arboreal; diurnal or nocturnal

REPRODUCTION
All species are oviparous, most producing two, or sometimes one, parchment-shelled eggs

DIET
Arthropods, primarily insects and spiders

GEKKONIDAE
WORLDWIDE COSMOPOLITAN GECKOS

The Gekkonidae is the largest of the seven families in the Gekkota, with 58 genera and over 1,300 species. It also contains eight of the ten largest gecko genera, and geckos with both dilated leaf-toes and slender, non-dilated, bent toes. All species are oviparous, laying one or two hard-shelled eggs, sometimes communally or often multi-clutching in a single year.

The most widespread lizard genus in the world is *Hemidactylus*, with 165 species. The most common species are the perianthropic house geckos, such as the Asian House Gecko (*H. frenatus*), Indo-Pacific House Gecko (*H. garnotii*), African House Gecko (*H. mabouia*), Mediterranean House Gecko (*H. turcicus*), and Flat-tailed House Gecko (*H. platyrurus*), which have established themselves across the tropical and subtropical world. The Indo-Pacific House Gecko is parthenogenetic, which makes it a very good colonizer, establishing new footholds, but a poor competitor when a sexual species like the Asian House Gecko arrives.

LEFT | The Indo-Pacific House Gecko (*Hemidactylus garnotii*) is parthenogenetic, which makes it an excellent colonizer. Note the calcium deposits in its neck, which will be used for egg formation.

RIGHT | The voracious and aggressive Tokay Gecko (*Gekko gecko*) is heard more often than it is seen, with its distinctive call of "to-kay."

DISTRIBUTION
Worldwide but poorly represented in the Americas and Australia

GENERA
Afroedura, Afrogecko, Agamura, Ailuronyx, Alsophylax, Altiphylax, Blaesodactylus, Bunopus, Calodactylodes, Chondrodactylus, Christinus, Cnemaspis, Crossobamon, Cryptactites, Cyrtodactylus, Cyrtopodion, Dixonius, Dravidogecko, Ebenavia, Elasmodactylus, Geckolepis, Gehyra, Gekko, Goggia, Hemidactylus, Hemiphyllodactylus, Heteronotia, Homopholis, Kolekanus, Lakigecko, Lepidodactylus, Luperosaurus, Lygodactylus, Matoatoa, Mediodactylus, Microgecko, Nactus, Narudasia, Pachydactylus, Paragehyra, Paroedura, Parsigecko, Perochirus, Phelsuma, Pseudoceramodactylus, Pseudogekko, Ptenopus, Ramigekko, Rhoptropella, Rhoptropus, Rhinogekko, Stenodactylus, Tenuidactylus, Trachydactylus, Trigonodactylus, Tropiocolotes, Urocotyledon, and Uroplatus

Many *Hemidactylus* exhibit more localized distributions in Asia, Africa, or Arabia, and although the majority of *Hemidactylus* originate from the Old World, there are three New World species, including the Guianan Leaf-toed Gecko (*H. palaichthus*).

Other widely distributed genera are found across Asia and the Indian and Pacific Oceans, such as the dwarf tree geckos (*Hemiphyllodactylus*), scaly-toed geckos (*Lepidodactylus*), and palm geckos (*Gekko*), or Australia, Asia, and the Pacific, such as the dtellas or four-clawed geckos (*Gehyra*). These genera contain widely distributed, colonizing, parthenogenetic species, including the Indo-Pacific Tree Gecko (*H. typus*), Mourning Gecko (*L. lugubris*), and Mutilated Gecko (*Gehyra mutilata*), which practices dermal autotomy to avoid predation.

HABITATS
Desert, rocky outcrops, rainforest, woodland, heathland, grassland, islands, and human structures

SIZE
SVL 1 in (23 mm) Broadley's Dwarf Gecko (*Lygodactylus broadleyi*) to 7¾ in (200 mm) Tokay Gecko (*Gekko gecko*) or Giant Leaf-tail Gecko (*Uroplatus giganteus*), the 8½ in (218 mm) Rodrigues Giant Gecko (*Phelsuma gigas*) is probably extinct

ACTIVITY
Arboreal, terrestrial, saxicolous, fossorial, or perianthropic; mostly nocturnal, but with diurnal genera, including *Cnemaspis*, *Lygodactylus*, *Phelsuma*, *Rhoptropella*, and *Rhoptropus*

REPRODUCTION
All species are oviparous, with clutches of two hard-shelled eggs; small species produce single eggs, but many species produce multiple clutches each year. Parthenogenetic species include the Mutilated Gecko (*Gehyra mutilata*), Indo-Pacific House Gecko (*Hemidactylus garnotii*), Indo-Pacific Tree Gecko (*Hemiphyllodactylus typus*), Mourning Gecko (*Lepidodactylus lugubris*), and Pelagic Gecko (*Nactus pelagicus*)

DIET
Arthropods; primarily insects and spiders, although larger species (*G. gecko*) prey on smaller geckos, while some species, such as the Ornate Day Gecko (*Phelsuma ornata*), also feed on nectar

GEKKONIDAE
AFRICAN GECKOS

Many of the endemic African geckos are confined to southern Africa, such as the diminutive African flat geckos (*Afroedura*), African leaf-toed geckos (*Afrogecko* and *Goggia*), African day geckos (*Rhoptropella* and *Rhoptropus*), Swartberg Rock Gecko (*Ramigekko swartbergensis*), and Namibian Festive Gecko (*Narudasia festiva*), all of which inhabit rocky areas. One of the most unusual habitats occupied by a gecko is the coastal saltmarshes near Port Elizabeth, South Africa, which are inhabited by Péringuey's Coastal Leaf-toed Gecko (*Cryptactites peringueyi*).

The savannas, rocky outcrops, and deserts of southern and eastern Africa also contain a diverse array of geckos, including the small thick-toed geckos (*Pachydactylus*) and larger tuberculate geckos (*Chondrodactylus* and *Elasmodactylus*). One of the largest

BELOW | The Small-scaled Leaf-toed Gecko (*Goggia microlepidota*), from the Western Cape, is one of many small rock-dwelling geckos from southern Africa.

ABOVE | When alarmed, the Giant Ground Gecko (*Chondrodactylus angulifer*) stands tall with its legs straight and its tail curved over its back like a scorpion, a posture from which it will hiss, lunge, and bite at its perceived enemy.

is the Giant Ground Gecko (*C. angulifer*), which may achieve a SVL of 4 in (100 mm) and inhabits burrows in gravel pans or dry river valleys. Noisy barking geckos (*Ptenopus*) also inhabit burrows, while large savanna trees are home to arboreal species, such as Wahlberg's Velvet Gecko (*Homopholis wahlbergii*). The tropical forests of East and West Africa are inhabited by four species of tail-pad geckos (*Urocotyledon*), with a fifth species, the Seychelles Surprise Gecko (*U. inexpectata*), curiously endemic to the Seychelles.

Several African gekkonid genera have ranges that extend far outside Africa. The African dwarf gecko genus *Lygodactylus* contains 68 species, mostly in the tropical forests of Central and West Africa, but species also inhabit Madagascar, Juan de Nova Island in the strait between Africa and Madagascar, and Brazil. The diurnal Asian genus *Cnemaspis* is also represented in the African rainforests, while the dwarf sand geckos (*Tropiocolotes*) occur across North Africa and Arabia.

GEKKONIDAE
ASIAN & AUSTRALASIAN GECKOS

The largest gecko genus (*Cyrtodactylus*) contains more than 300 species of bent-toed geckos distributed from Kashmir to the Solomon Islands, with new species being regularly described. They are slender-toed rock-dwellers, common on limestone karst, and related to the Mediterranean Bent-toed Gecko (*Mediodactylus kotschyi*) that enters southeastern Europe. Bent-toed geckos are nocturnal and their rocky outcrops, in South and Southeast Asia, may be occupied diurnally by the Asian day geckos (*Cnemaspis*). The island bent-toed geckos (*Nactus*) occur in New Guinea, Queensland, and the Pacific, and include the parthenogenetic Pelagic Gecko (*N. pelagicus*).

Western Asia and the Arabian Peninsula also contain numerous arid-habitat geckos, such as the long-legged spider geckos (*Agamura* and *Rhinogekko*), Asian tuberculate geckos (*Bunopus*), short-fingered sand geckos (*Stenodactylus*), and Asian dwarf geckos (*Microgecko*), which are among the smallest members of the Gekkonidae. Some geckos exhibit extremely localized distributions, including Ziaie's Pars Gecko (*Parsigecko ziaiei*) from Hormozgan, Iran, and the Golden Gecko (*Calodactylodes illingworthorum*), endemic to the Nuvara Gala Rock, Sri Lanka.

BELOW | The Chat-tan Cave Bent-toed Gecko (*Cyrtodactylus chanhomeae*) is a member of the largest gecko genus with 307 species currently described.

Geckos also inhabit the rainforests of Southeast Asia. Among the best known are the cryptically patterned flying geckos (*Ptychozoon*, now a subgenus of *Gekko*), which can leap into the air and glide to safety when they feel threatened, while the Philippines are home to two endemic genera, the wolf geckos (*Luperosaurus*) and Philippine geckos (*Pseudogekko*). The Australian southern geckos (*Christinus*) comprise two species from mainland Australia, and one (*C. guentheri*) from Norfolk and Lord Howe islands in the Tasman Sea between Australia and New Zealand, while five species of prickly geckos (*Heteronotia*) also inhabit Australia, and three species of Pacific geckos (*Perochirus*) occur in Micronesia and Vanuatu.

ABOVE | Flying geckos (*Ptychozoon*, now *Gekko*) are able to glide away from predators due to fringes of skin along their body and webbing between their toes (see inset) that catch the air and slow their descent. They are not flying, they are parachuting.

GEKKONIDAE
INDIAN OCEAN GECKOS

The islands of the Indian Ocean also possess a very rich gecko fauna. The best-known and most visible geckos are the 50 species of stunning day geckos (*Phelsuma*) with their resplendent array of greens, blues, and reds. These geckos are important pollinators of island plants, but sadly the two Rodrigues Island species are already extinct.

While the day geckos boldly advertise their presence, other geckos have taken camouflage to the extreme, not only adopting moss- or lichen-like patterning but also cryptic body shapes. Among these secretive nocturnal geckos are the Madagascan flat-tailed geckos (*Uroplatus*), 19 species that so resemble the mottled branches on which they shelter that, when immobile, they are almost invisible. Some species have frills along their flanks and limbs, to break up their body shape, and several have flattened tails that mimic dead leaves so perfectly that they even possess ragged edges that resemble insect damage. Other species resemble yellowish blades of grass.

Among the strangest of Indian Ocean geckos are the Madagascan and Comoros fish-scaled geckos (*Geckolepis*), which have bodies covered in large cycloid scales that really do resemble the scales of fish. Many lizards autotomize their tails when caught by a predator, but these geckos autotomize whole areas of scaly skin (see page 74). The three species of endemic Seychelles bronze geckos (*Ailuronyx*) also autotomize their skin.

One of the remotest Indian Ocean locations is the Socotran Archipelago, between Yemen and Somalia, and it is here that *Hemidactylus* has speciated greatly, with ten species present, six of them endemic.

ABOVE LEFT | The toes of the Southern Flat-tail Gecko (*Uroplatus sikorae*) exhibit webbing that helps to break up the gecko's outline, even at this magnification.

LEFT | The Satanic Leaf-tailed Gecko (*Uroplatus phantasticus*) has tiny horns over its eyes and a body and tail that resemble dead leaves, even down to the leaf's veins.

RIGHT | The day-glo patterned Lined Day Gecko (*Phelsuma lineata*), which is named for its broad, dark, lateral stripe, is a widely distributed Madagascan species with four subspecies.

EUBLEPHARIDAE—AELUROSCALABOTINAE & EUBLEPHARINAE
EYELID GECKOS

Most geckos lack eyelids, instead possessing transparent spectacles over the eyes, but geckos in the Eublepharidae do possess eyelids and can blink. They also lack dilated toe pads, a sign of their generally more terrestrial existence.

There are two subfamilies, the Eublepharinae and Aeluroscalabotinae, the latter with a single species, the Cat Gecko (*Aeluroscalabotes felinus*). Unlike the eublepharines, this is a semi-arboreal species, found in lowland rainforests and peat forests in Borneo and Peninsular Malaysia. It is associated with low vegetation or fallen trees.

The Eublepharinae contains five genera and 39 terrestrial species that curiously wave their bulbous tails before grasping their arthropod prey. The banded geckos (*Coleonyx*) occur from southwestern USA to Panama, in deserts, dry forests, and rocky habitats—they are the only American eublepharids. Two genera occur in Africa, the East African clawed geckos (*Holodactylus*) and the African fat-tailed geckos (*Hemitheconyx*), which includes both East and West African species. Both genera are adapted for life in sandy savanna or dry forest habitats and they dwell in subterranean burrows.

The West Asian leopard geckos (*Eublepharis*), which occur from Syria to eastern India, are desert- or semidesert-dwellers that also spend their days in burrows, emerging at night to hunt. There are six species, with the Common Leopard Gecko (*E. macularius*) one of the most popular pet-trade lizards. The largest eublepharid genus is *Goniurosaurus*, which currently contains 22 species. These East Asian leopard geckos are concentrated in two clusters, in northern Vietnam and southern China, including Hainan Island, and the Ryukyu Islands of Japan. Unfortunately, when new species are described they become the focus of illegal collecting for the pet trade, and island species are particularly vulnerable to exploitation.

AELUROSCALABOTINAE

DISTRIBUTION (BLUE ON MAP)
Southeast Asia

GENUS
Aeluroscalabotes

HABITATS
Lowland rainforest and peat forest to 3,280 ft (1,000 m)

SIZE
SVL 4¾ in (122 mm) Cat Gecko (*Aeluroscalabotes felinus*)

ACTIVITY
Semi-arboreal; nocturnal

REPRODUCTION
Oviparous, producing 1–2 parchment-shelled eggs

DIET
Arthropods, primarily crickets and cockroaches

LEFT | The Cat Gecko (*Aeluroscalabotes felinus*) is a gracile Southeast Asian rainforest and peat forest gecko that blends in well with its surroundings.

ABOVE | Taylor's Fat-tailed Gecko (*Hemitheconyx taylori*) is a nocturnal semidesert-dwelling species from Somalia and eastern Ethiopia.

RIGHT | The Central American Banded Gecko (*Coleonyx mitratus*) is the southernmost American eyelid gecko.

EUBLEPHARINAE

DISTRIBUTION (RED ON MAP)
East, Southeast, and Central Asia, East and West Africa, and North America

GENERA
Coleonyx, *Eublepharis*, *Goniurosaurus*, *Hemitheconyx*, and *Holodactylus*

HABITATS
Stony or sandy desert and semidesert, limestone karst, and islands

SIZE
SVL 2⅔ in (67 mm) Texas Banded Gecko (*Coleonyx brevis*) to 10 in (252 mm) Western Indian Leopard Gecko (*Eublepharis fuscus*)

ACTIVITY
Terrestrial, saxicolous, or semi-fossorial; all species are nocturnal

REPRODUCTION
All species are oviparous, producing 1–2 parchment-shelled eggs, and often multi-clutching

DIET
Arthropods, primarily insects and spiders

PHYLLODACTYLIDAE
OLD WORLD LEAF-TOED GECKOS

Phyllodactylidae is a cryptic transatlantic clade that was only distinguished and separated from Gekkonidae in 2008. Although member species are commonly referred to as the leaf-toed geckos, there are geckos in other families that also possess leaf-toes, and not all phyllodactylid geckos exhibit them. Phyllodactylidae was identified using molecular analyses rather than morphological characteristics.

LEFT | A Dragon-tree Gecko (*Haemodracon riebeckii*) on a young Desert Rose (*Adenium obesum socotranum*), both species being Socotran endemics.

BELOW | The genus *Haemodracon* is endemic to Socotra and contains two species. This is a close-up of the head of *H. riebeckii*.

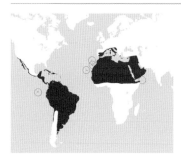

DISTRIBUTION
Central and South America, West Indies, Europe, North Africa, Middle East, Madeira, Canary and Cape Verde islands, and Socotra

GENERA
Asaccus, Garthia, Gymnodactylus, Haemodracon, Homonota, Phyllodactylus, Phyllopezus, Ptyodactylus, Tarentola, and *Thecadactylus*

HABITATS
Rainforest, dry forest, Cerrado, Caatinga, rocky outcrops, islands, and abandoned buildings

SIZE
SVL 1⅓ in (33 mm) Coquimbo Marked Gecko (*Garthia penai*) to 6 in (155 mm) Giant Wall Gecko (*Tarentola gigas*)

ACTIVITY
Terrestrial, arboreal, or saxicolous; primarily nocturnal, but some species are crepuscular or diurnal (*Tarentola*)

Four genera are distributed across Europe, Africa, and Western Asia. The Middle Eastern leaf-toed geckos (*Asaccus*) of Arabia, Turkey, and Iran, and the fan-fingered geckos (*Ptyodactylus*) from Arabia and North and West Africa, have long, slender legs and toes that terminate in broad, distinctive toe pads. They are nocturnal, saxicolous geckos that inhabit rock faces, caves, and human structures such as old forts.

The Socotran Archipelago is home to an endemic genus, *Haemodracon*, containing two species that inhabit the distinctive red-sapped Dragon Blood Trees (*Dracaena cinnabari*). The Dragon-tree Gecko (*H. riebeckii*) is even reported to eat the fruit of the tree.

Southern Europe and North Africa are home to the wall geckos (*Tarentola*). Most are rock-dwellers that also inhabit the walls of human habitations, such as the Moorish Gecko (*T. mauritanica*), but some species are terrestrial in habit, including the large-headed, squat-bodied Helmeted Gecko (*T. chazaliae*) from northwest Africa. Many species are found on islands in the Atlantic, including Madeira (*T. mauritanica*) and the Selvagens (*T. bischoffi*), and the genus has speciated greatly on the Canary Islands (four species) and Cape Verde Islands (13 species).

REPRODUCTION
All species are oviparous, usually producing 1–2 hard-shelled eggs, depending on the size of the species

DIET
Arthropods, primarily insects, although the Dragon-tree Gecko (*Haemodracon riebeckii*) may eat the fruit of the Dragon Blood Tree (*Dracaena cinnabari*)

ABOVE | The Helmeted Gecko (*Tarentola chazaliae*) from the desert of northwest Africa is so named because of its seemingly armored head.

PHYLLODACTYLIDAE
NEW WORLD LEAF-TOED GECKOS

Not all *Tarentola* are Old World; three species are found in the Caribbean, the American Wall Gecko (*T. americana*), Jamaican Wall Gecko (*T. albertschwartzi*), and Cuban Wall Gecko (*T. crombiei*).

The nominate genus of the family is *Phyllodactylus*, and it is also the largest, with 59 species. These occur over a vast area, from southern California to southern Argentina, in every mainland country except Canada, and on most Caribbean islands, but only one species, the Peninsula Leaf-toed Gecko (*P. nocticolus*), enters the United States. Twelve species occur on the Galapagos Islands, all but two of them endemic to the archipelago. Most *Phyllodactylus* are small and discrete, and all have expanded leaf-toes.

LEFT | The Galapagos Leaf-toed Gecko (*Phyllodactylus galapagensis*) is one of ten species from the genus that are endemic to the famous Ecuadorian archipelago.

ABOVE | The Giant Turnip-tailed Gecko (*Thecadactylus rapicauda*) is a large, noisy, and much-feared denizen of the neotropical rainforests.

The aptly named turnip-tailed geckos (*Thecadactylus*) are also quite widely distributed, from the Yucatán Peninsula, Mexico, to the Lesser Antilles and the Amazon Basin. These large geckos are much feared as portents of evil in Central America, but fear of geckos calling at night is not uncommon in the tropics.

The remaining phyllodactylid genera are South American. Two genera of marked geckos (*Garthia* and *Homonota*) inhabit southern South America. *Garthia* species occur west of the Andes and possess claws that can be retracted into the toe sheath, while *Homonota* occur east of the Andes and have non-retractable claws.

Six species of South American leaf-toed geckos (*Phyllopezus*) occur in South America, in arid Caatinga, Cerrado, Chaco, or dry forest habitats. They may be terrestrial or arboreal, living on spiny xerophytic trees. The Brazilian naked-toed geckos (*Gymnodactylus*), which are endemic to Brazil, are not leaf-toed, but possess slender bent toes like the genus *Cyrtodactylus* (Gekkonidae). They are also terrestrial inhabitants of arid Cerrado or Caatinga habitats.

SPHAERODACTYLIDAE
NEW WORLD DWARF & LEAST GECKOS

The Sphaerodactylidae is a family of generally small dwarf or least geckos that spans the Atlantic with New and Old World genera, but the majority of species occur in the Americas, including the largest genus, *Sphaerodactylus*, which gives the family its name. This is the fourth largest gecko genus, with 107 species.

Sphaerodactylus are small geckos, most having a SVL of 1–1⅓ in (26–34 mm), but they can be even smaller. The smallest gecko in the world is the Marche Leon Least Gecko (*S. elasmorhynchus*) from Haiti, Hispaniola, with a SVL of ⅔ in (17 mm), but several other species are close rivals. The genus occurs from southern Mexico to Colombia, including the Pacific island of Gorgona, and extreme northwestern Ecuador, and from southern Florida, through the Greater and Lesser Antilles to Trinidad, northeastern Venezuela and Guyana. It does not extend far into South America. In some English-speaking West Indian countries these small geckos are known as "woodslaves."

The Caribbean is also home to the genus *Aristelliger*, the croaking geckos, while the dwarf bent-toed geckos (*Gonatodes*) occur from southern Mexico to the Caribbean and throughout northern South America. Both these genera may be terrestrial or arboreal, and active by day or night. The Latin

BELOW | The Amazonian Leaf-litter Gecko (*Chatogekko amazonicus*) is a tiny lizard, seen here on a finger, but this is not the smallest sphaerodactylid gecko by a considerable margin.

DISTRIBUTION
Central and South America, West Indies, northwest and eastern Africa, southern Europe, Arabia, and western and Central Asia

GENERA
Aristelliger, Chatogekko, Coleodactylus, Euleptes, Gonatodes, Lepidoblepharis, Pristurus, Pseudogonatodes, Quedenfeldtia, Saurodactylus, Sphaerodactylus, and Teratoscincus

HABITATS
Rainforest, plantations, coastal woodland, sandy desert, stony semidesert and mountains, islands, and under human trash

SIZE
SVL ⅔ in (17 mm) Marche Leon Least Gecko (*Sphaerodactylus elasmorhynchus*) to 6 in (150 mm) Spotted Croaking Gecko (*Aristelliger lar*)

ACTIVITY
Terrestrial, arboreal, saxicolous, or semi-

American mainland, particularly Amazonia, is home to four genera of diminutive terrestrial geckos: the leaf-litter geckos *Chatogekko* and *Coleodactylus*, and the sheath-clawed geckos *Lepidoblepharis* and *Pseudogonatodes*. All these small geckos are seen as prey by larger lizards, snakes, mammals, birds, frogs, and even large invertebrates.

ABOVE | The Yellow-headed Gecko (*Gonatodes albogularis*) is widely distributed in the Caribbean, Central America, and northern South America, and introduced to Florida. It is sexually dichromatic, only adult males possessing the yellow or orange head.

BELOW | The Ashy Gecko (*Sphaerodactylus elegans*) is a dwarf gecko species native to Cuba and Hispaniola, and introduced to Florida, where they have established a thriving population.

fossorial; primarily nocturnal, but some genera are crepuscular (*Saurodactylus*) or diurnal (*Gonatodes, Pristurus, Quedenfeldtia*)

REPRODUCTION
All species are oviparous, usually producing single hard-shelled eggs, or two eggs in large species (*Teratoscincus*)

DIET
Arthropods; primarily insects, from ants to beetles, or sometimes arachnids, but some species (*Aristelliger*) also eat fruit

SPHAERODACTYLIDAE
OLD WORLD DWARF GECKOS

Five widely distributed genera occur in the Old World. The sole European representative is the European Leaf-toed Gecko (*Euleptes europaea*), found on Corsica and Sardinia, with toeholds on the French and Italian mainlands, and some Tunisian islands. Europe's smallest gecko, it has a SVL of 1½ in (40 mm). It is nocturnal and saxicolous, living on rock faces or buildings.

Two genera occur in northwestern Africa. The lizard-fingered geckos (*Saurodactylus*) inhabit arid coastal scrub or stony deserts, from Algeria to Western Sahara, possibly as far south as Mauritania, while the Atlas day geckos (*Quedenfeldtia*) occur in the Moroccan Atlas Mountains. Both genera are diurnal, *Quedenfeldtia* because the nights at high elevations are cold. *Saurodactylus* geckos avoid the hottest part of the day and are crepuscular in summer.

Northeast Africa, the Socotran Archipelago, and Arabia are home to 26 species of semaphore geckos (*Pristurus*), although the Adrar Atar Semaphore Gecko (*P. adrarensis*) occurs 2,920 miles (4,700 km) west in Mauritania. The common name refers to the territorial tail-waving of the males. *Pristurus* means "saw-tailed," a reference to the serrated dorsal tail keel of some species.

LEFT | Carter's Semaphore Gecko (*Pristurus carteri*) is sometimes called a Scorpion-tailed Gecko because of the way it curves its serrated tail over its head in defense.

ABOVE | The Persian Wonder Gecko (*Teratoscincus keyserlingii*) is one of the largest sphaerodactylid geckos. It has huge eyes, which are probably the origin of the common name "wonder gecko."

Among the largest sphaerodactylids are the Central and Western Asian wonder geckos (*Teratoscincus*), from *terato*, meaning "wonder." These nocturnal desert geckos may exceed a SVL of 4 in (100 mm), and have large heads and very large eyes. The largest species, the Persian Wonder Gecko (*T. keyserlingii*), also occurs in the UAE, Oman, and Qatar. It uses a number of defensive tactics, from dermal autotomy to standing with legs stiff, moving the tail back and forth, or produce a rustling sound, or leaping forward to bite. Most sphaerodactylids lay single eggs, but *Teratoscincus* lays two.

ABOVE | Europe's smallest gecko, and also its only sphaerodactylid species, the European Leaf-toed Gecko (*Euleptes europaea*) is a Mediterranean island species that only occurs on the mainland in small coastal parts of France and Italy.

LEFT | A juvenile Emerald Tree Skink (*Lamprolepis smaragdina*) is highly agile and well camouflaged for life aloft.

INFRAORDER SCINCOMORPHA

The infraorder Scincomorpha contains four families. Three of these exhibit relatively localized distributions: the Cordylidae (girdled and flat lizards) of Africa; Gerrhosauridae (plated lizards) of Africa and Madagascar; and Xantusiidae (night lizards) from North and Central America, and Cuba.

The fourth family is the Scincidae (skinks), which contains 24 percent of all lizards and is distributed worldwide, although very few species occur above latitude 60 degrees north. Many authors divide the Scincidae into seven subfamilies, but these are treated as families by others.

Of the seven subfamilies, the Acontinae (lance skinks) is southern African; the Egerniinae (social skinks) is Australo-Melanesian; the Lygosominae (supple, writhing, and tree skinks) is Afro-Asian; the Mabuyinae (sun and ground skinks) and Scincinae (burrowing skinks) are Afro-Asian and American, the former being more southern hemispheric and the latter more northern hemispheric; the Eugongylinae (austral and snake-eyed skinks) is Afro-Asian, Australasian, and Oceanic; while the Sphenomorphinae (Australo-Asian skinks) is Asian, Australasian, Oceanic, and American.

CORDYLIDAE—CORDYLINAE & PLATYSAURINAE
GIRDLED & FLAT LIZARDS

The Cordylinae (girdled lizards) contains nine genera, the largest being the small to medium-sized girdled lizards (*Cordylus*), named for their strongly keeled scales. Most species inhabit rocky outcrops, but three are arboreal tree- or euphorbia-dwellers. Largest are the dragon lizards (*Smaug*), named for Tolkien's dragon from *The Hobbit*. Many, including the largest cordylid, the Sungazer (*S. giganteus*), are extremely rugose, with spiny whorls around the tail, but some species are less rugose, such as the Gorongosa Girdled Lizard (*S. mossambicus*).

Also less spiny, the two species of cliff lizards (*Hemicordylus*) are melanistic because they dwell in the cool montane habitats on the Cape of South Africa. The Blue-spotted Lizard (*Ninurta coeruleopunctatus*) and the crag lizards (*Pseudocordylus*) are covered in small,

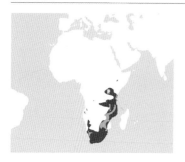

CORDYLINAE

DISTRIBUTION (RED AND ORANGE ON MAP)
Southern and East Africa

GENERA
Chamaesaura, Cordylus, Hemicordylus, Karusasaurus, Namazonurus, Ninurta, Ouroborus, Pseudocordylus, and *Smaug*

HABITATS
Limestone or sandstone outcrops, karoo, semidesert, coastal fynbos, montane grassland, and acacia woodland

SIZE
SVL 2⅔ in (67 mm) Eastern Dwarf Girdled Lizard (*Cordylus aridus*) to 8 in (205 mm) Sungazer (*Smaug giganteus*)

ACTIVITY
Terrestrial, saxicolous, or arboreal; all species are diurnal

LEFT | This Armadillo lizard (*Ouroborus cataphractus*) wears a suit of spiny armor and, rolled into a ball, would make a very unappealing meal for a predator.

ABOVE | This gaudy Namaqua flat lizard (*Platysaurus capensis*) basks on a rock and appears vulnerable to predators, but at any sign of danger it will vanish into a nearby crevice.

weakly keeled scales. Both genera inhabit cool, wet, coastal fynbos, basking on rocky outcrops, even in cold weather. Two species of karoo lizards (*Karusasaurus*) and the Namaqualand girdled lizards (*Namazonurus*) inhabit low-lying semiarid habitats in Namibia and South Africa.

Ouroboros was a mythical dragon that held its own tail in its mouth, an apt generic name for the excessively spiny Armadillo Lizard (*Ouroborus cataphractus*) which, if prevented from escaping, will roll into a ball, tuck its legs in, and grasp its tail in its jaws, presumably making it a difficult meal to swallow.

The strangest cordylids are the grass lizards (*Chamaesaura*, see page 21). These extremely elongate, long-tailed lizards have hind limbs reduced to small flaps, and forelimbs also reduced, or absent. This body form is perfect for "swimming" through the savanna grasslands.

The Platysaurinae contains the flat lizards (*Platysaurus*), 16 species of brightly colored, dorsoventrally compressed rock-dwellers, which have small-scaled, velvety bodies, but more rugosely keeled tails.

REPRODUCTION
All species are viviparous, giving birth to litters of 1–4 or more rarely 5–7 (*Pseudocordylus, Smaug*) or 8–12 (*Chamaesaura*) neonates, depending on species and size

DIET
Arthropods, primarily grasshoppers, beetles, and spiders, and occasionally vegetation; some large species take small vertebrates

PLATYSAURINAE

DISTRIBUTION (ORANGE ON MAP)
Southern and East Africa

GENUS
Platysaurus

HABITATS
Granite and sandstone outcrops, savanna, and woodland

SIZE
SVL 2 in (52 mm) Mitchell's Flat Lizard (*Platysaurus mitchelli*) to 5¾ in (146 mm) Emperor Flat Lizard (*P. imperator*)

ACTIVITY
Arboreal, terrestrial, or saxicolous; all species are diurnal

REPRODUCTION
All species are oviparous, producing clutches of two eggs

DIET
Arthropods, primarily flies, beetles and their larvae, although some species eat flowers, fruit, leaves, and seeds

GERRHOSAURIDAE—GERRHOSAURINAE & ZONOSAURINAE
PLATED LIZARDS

LEFT | The Rough-scaled Plated Lizard (*Broadleysaurus major*) is protected by a coat of spiny armor while other plated lizards may have smoother scales. This specimen still has its juvenile patterning.

The African subfamily Gerrhosaurinae (African plated lizards and seps) contains five genera. The smallest is the Dwarf Plated Lizard (*Cordylosaurus subtessellatus*), which inhabits sandy areas of southwestern Africa, and possesses a smooth-scaled, black body with bold dorsolateral cream stripes pointing toward a brilliant blue tail, for distracting the attention of potential predators. It resembles a skink more than a plated lizard.

Other plated lizards are more robust and possess regular rings of square, plate-like scales resembling armor. Most belong to the typical plated lizard genus *Gerrhosaurus*, but the giant plated lizards are now placed in a separate genus (*Matobosaurus*), while the most rugose species, the Rough-scaled Plated Lizard (*Broadleysaurus major*), is also placed in a separate genus. A distinctive characteristic of all plated lizards is a deep lateral fold that runs along the flanks.

CORDYLINAE

DISTRIBUTION (RED ON MAP)
Southern and East Africa

GENERA
Broadleysaurus, Cordylosaurus, Gerrhosaurus, Matobosaurus, and *Tetradactylus*

HABITATS
Upland savanna, bushveld, desert sand dunes, coastal forest, and fynbos

SIZE
SVL 2¼ in (55 mm) Dwarf Plated Lizard (*Cordylosaurus subtessellatus*) to 11¼ in (285 mm) Common Giant Plated Lizard (*Matobosaurus validus*)

ACTIVITY
Arboreal, terrestrial, or saxicolous; all species are diurnal

REPRODUCTION
All species are oviparous, laying clutches of 2–12 leathery-shelled eggs, depending on the species

DIET
Arthropods, from termites to grasshoppers and scorpions; also fruit, flowers, and seeds, with the desert-dwelling Nara plant an important food for the

Body elongation and limb reduction are also evident in the gerrhosaurids, in the seps (*Tetradactylus*), which inhabit grasslands and coastal fynbos. While some species possess well-developed but short legs, others, such as the African Long-tailed Seps (*T. africanus*), lack any forelimbs. Some species are vulnerable to habitat change, and Eastwood's Long-tailed Seps (*T. eastwoodae*) is believed extinct.

The subfamily Zonosaurinae is endemic to the Indian Ocean islands and contains two genera. The Madagascan girdled lizards (*Zonosaurus*) are smooth-scaled and closely resemble African plated lizards. Although primarily terrestrial, two green species are arboreal, Boettger's Girdled Lizard (*Z. boettgeri*) and the Dark-spotted Girdled Lizard (*Z. maramaintso*). The Common Girdled Lizard (*Z. madagascariensis*) has been introduced to Glorioso Island and Cosmoledo Atoll, and Aldabra Atoll in the southern Seychelles. Two species of terrestrial keel-scaled plated lizards (*Tracheloptychus*) also inhabit Madagascar.

ABOVE LEFT | The Giant Plated Lizard (*Matabosaurus validus*) is a large, robust, smooth scaled species.

LEFT | Harold Meier's Girdled Lizard (*Zonosaurus haroldmeieri*) is a terrestrial inhabitant of forest edges in northern Madagascar.

Desert Plated Lizard (*Gerrhosaurus skoogi*). Small vertebrates including lizards and small tortoises are taken by large *M. validus*

ZONOSAURINAE

DISTRIBUTION (ORANGE ON MAP)
Madagascar, Seychelles, and Glorioso and Cosmoledo islands

GENERA
Tracheloptychus and *Zonosaurus*

HABITATS
Rainforest, montane forest, riverine forest, dry forest on limestone, thorn forest, and plantations

SIZE
SVL 2¾ in (70 mm) Bronze Girdled Lizard (*Zonosaurus aeneus*) to 9⅔ in (246 mm) Southeastern Girdled Lizard (*Z. maximus*)

ACTIVITY
Terrestrial and arboreal; all species are diurnal

REPRODUCTION
All species are oviparous, producing clutches of up to eight eggs

DIET
Arthropods, primarily insects and spiders, but some *Zonosaurus* also feed on vegetation

XANTUSIIDAE—CRICOSAURINAE, LEPIDOPHYMINAE & XANTUSIINAE
NIGHT LIZARDS

The Xantusiidae is a small neotropical family of small, slender, secretive lizards with granular body scales, large regular head scutes, and gecko-like spectacles over their lidless eyes.

Xantusiidae contains three subfamilies. The Cricosaurinae is a monotypic subfamily containing only the Cuban Night Lizard (*Cricosaura typica*), which appears to be confined to southeastern Cuba. It is found in dry woodland, sheltering under stones and bark or in the loose subsoil during the day, becoming active at night. It moves in a serpentine fashion because its short limbs are ineffective for locomotion.

The tropical night lizards (*Lepidophyma*) comprise the second subfamily, Lepidophyminae. Twenty species occur from northeastern Mexico to Panama. The most widely distributed species is the Yellow-spotted Night Lizard (*L. flavimaculatum*),

LEFT | The Yellow-spotted Night Lizard (*Lepidophyma flavomaculatum*) is distributed from Mexico to Panama, the Panama population being parthenogenetic.

RIGHT | Resembling a European lacertid lizard, the Desert Night Lizard (*Xantusia virgilis*) of southwest USA and northwest Mexico belies its name by being largely diurnal or crepuscular.

CRICOSAURINAE	**ACTIVITY**	**LEPIDOPHYMINAE**
DISTRIBUTION (BLUE ON MAP) Cuba	Semi-fossorial; nocturnal	**DISTRIBUTION (ORANGE ON MAP)** Mexico and Central America
GENUS *Cricosaura*	**REPRODUCTION** Not known; presumed viviparous	**GENUS** *Lepidophyma*
HABITATS Thorn forest and cactus scrub on karst limestone	**DIET** Small arthropods and their larvae	**HABITATS** Rainforest, in decaying tree trunks or rock piles
SIZE SVL 1½ in (40 mm) Cuban Night Lizard (*Cricosaura typica*)		**SIZE** SVL 1½ in (37 mm) Liner's

which occupies almost the entire generic range along the Caribbean versant. Smith's Tropical Night Lizard (*L. smithii*) occurs from Mexico to El Salvador, on the Pacific versant, but most other species exhibit localized distributions. All species are viviparous, and some comprise all-female parthenogenetic species, such as the Costa Rican Night Lizard (*L. reticulatum*).

The nominate subfamily Xantusiinae (northern night lizards) also contains a single genus (*Xantusia*), with 14 species distributed across southwestern USA and northwestern Mexico. They differ from both tropical and Cuban night lizards in possessing vertically elliptical, rather than round, pupils. Some also belie their common names by being diurnal under suitable conditions. Most, such as the Granite Night Lizard (*X. henshawi*), are sedentary, sheltering in rocky crevices or under rocks. Some species, including the Desert Night Lizard (*X. vigilis*), are associated with succulent vegetation, such as yucca or agave, and are more arboreal. The Island Night Lizard (*X. riversiana*) is more generalist in its habitat preferences, being found living on prickly pears cactus and under beach driftwood.

Tropical Night Lizard (*Lepidophyma lineri*) to 6 in (153 mm) Yellow-spotted Tropical Night Lizard (*L. flavimaculatum*)

ACTIVITY
Terrestrial, arboreal, or saxicolous; crepuscular or nocturnal

REPRODUCTION
Viviparous, with litters of 2–5, occasionally 8. Some populations are parthenogenetic

DIET
Arthropods, primarily insects and spiders

XANTUSIINAE

DISTRIBUTION (RED ON MAP)
North America

GENUS
Xantusia

HABITATS
Semiarid rocky canyons, chaparral, woodland, and upland desert; under exfoliated rocks, leaves, or bark

SIZE
SVL 1½ in (36 mm) Sherbrooke's Night Lizard (*Xantusia sherbrookei*) to 4½ in (117 mm) Island Night Lizard (*X. riversiana*)

ACTIVITY
Terrestrial, arboreal, or saxicolous; nocturnal or diurnal

REPRODUCTION
Viviparous, with litters of 1–7 neonates depending on species

DIET
Arthropods, primarily insects or their larvae in rotting vegetation, as well as spiders and centipedes; also flowers and seeds

SCINCIDAE—ACONTINAE
LANCE SKINKS

The huge Scincidae (skinks) has been split into smaller groups on numerous occasions in the past. The arrangement used here recognizes seven separate clades but retains them as subfamilies within the Scincidae. However, the author is aware of a recent paper that elevated all these subfamilies to family status, and described two further small families, the Ateuchosauridae and the Ristellidae, which are included here in the Scincinae and Sphenomorphinae, respectively.

The subfamily Acontinae comprises two genera and 30 species confined to sub-Saharan Africa. They share a number of characteristics, notably a divided frontal bone to the skull, a total lack of limbs and external ear openings, and a short, stumpy tail—equivalent to less than 22 percent of their total length. They are variously referred to as legless skinks, dart skinks, or lance skinks.

Lance skinks are fossorial, mostly inhabiting sandy soils, where they feed on beetle larvae,

ACONTINAE

DISTRIBUTION
Sub-Saharan Africa

GENERA
Acontias and *Typhlosaurus*

HABITATS
Coastal scrub, forest and dunes, dry woodland, thicket, grassland, and semidesert

SIZE
SVL 5 in (126 mm) Coastal Dwarf Lance Skink (*Acontias litoralis*) to 19⅔ in (500 mm) Giant Lance Skink (*A. plumbeus*)

ACTIVITY
Fossorial; presumed nocturnal

REPRODUCTION
Viviparous, with litters of 1–4 neonates, or up to 14 in large species (*A. plumbeus*)

LEFT | Gray's Lance Skink (*Acontias grayi*) inhabits the sandy soils of the coastal fynbos of Western Cape Province, but it can be found up to 2,950 ft (900 m) above sea level.

ABOVE | The Pink Blind Lance Skink (*Typhlosaurus vermis*) inhabits coastal sandveld habitats in Namaqualand, Northern Cape Province, up to the Namibian border. Its eyes are reduced to small pigmented areas under translucent scales.

DIET
Small arthropods, including termites, beetle larvae, and centipedes; also earthworms

termites, centipedes, or earthworms. All species are believed to be viviparous, producing 1–14 neonates, depending on the species and the female's size. The genus *Acontias* consists of 24 southern African lance skinks, and one East African species, Percival's Lance Skink (*Acontias percivali*), which occurs on the Kenya–Tanzania border.

The remaining five species are the blind lance skinks (*Typhlosaurus*), five extremely slender species with vestigial eyespots under translucent head scales, rather than functional eyes. The genus is confined to Namaqualand and neighboring areas in South Africa and southern Namibia.

Most lance skinks are small, measuring less than 7¾ in (200 mm) SVL, but the Giant Lance Skink (*A. plumbeus*) from southwestern Africa may reach almost 20 in (500 mm).

SCINCIDAE—SCINCINAE
BURROWING SKINKS

The subfamily Scincinae, including the Ateuchosauridae, contains 35 genera and almost 300 species that occur through southern Europe, Africa, the Middle East, Asia, North and Central America, and across the Indian Ocean. Many species illustrate a degree of body elongation and limb and digit reduction due to their burrowing lifestyles, such as the cylindrical skinks (*Chalcides*), and some are completely fossorial and limbless, such as the African legless skinks (*Melanoseps* and *Feylinia*).

The blue-tailed skinks of genus *Plestiodon* demonstrate an interesting distribution, being found in eastern Asia and North America. All the Asian species are oviparous, some being excellent mothers that guard their eggs, but approximately half of the American species are viviparous. This genus also contains the only skink to occur on Bermuda, the Bermuda Skink (*P. longirostris*).

One of the most endangered species is the Mascarene skink genus *Gongylomorphus*, which went

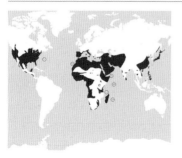

SCINCINAE

DISTRIBUTION
North and Central America, southern Europe, Africa, Arabia, Asia, Madagascar, Canary Islands, Seychelles, Mascarenes, and Socotra

GENERA
Amphiglossus, Ateuchosaurus, Barkudia, Brachymeles, Brachyseps, Chalcides, Chalcidoseps, Eumeces, Eurylepis, Feylinia, Flexiseps, Gongylomorphus, Grandidierina, Hakaria, Janetaescincus, Jarujinia, Madascincus, Melanoseps, Mesoscincus, Nessia, Ophiomorus, Pamelaescincus, Paracontias, Plestiodon, Proscelotes, Pseudoacontias, Pygomeles, Scelotes, Scincopus, Scincus, Scolecoseps, Sepsina, Sepsophis, Typhlacontias, and *Voeltzkowia*

LEFT | The shovel-snouted Eastern Sandfish (*Scincus mitranus*) from the Arabian Peninsula and Iran, is perfectly built for life on loose sand dunes. It basks during the day but if it feels threatened it will dive into the sand and swim deeply into the dune like a fish through water.

extinct on Réunion Island in the mid nineteenth century, probably due to the arrival of invasive rats, cats, and the lizard-eating Island Wolfsnake (*Lycodon capucinus*). It has now been extirpated from most of Mauritius, except the extreme southwest and a few small offshore islands, for the same reasons. Several other genera are confined to Indian Ocean islands, *Hakaria* on Socotra, *Janetaescincus* and *Pamelaescincus* in the Seychelles, and *Flexiseps* on the Comoro Islands and Madagascar.

The sandfish (*Scincus*) are sand-dune skinks from North Africa and Arabia that have flattened, shovel-like snouts. They are well named because when threatened they rapidly burrow and "swim" down through the loose sand.

LEFT | Juvenile Western Skinks (*Plestiodon skiltonianus*) have bright blue tails to attract the attention of predators, including cannibalistic males, away from the head end. This is an oviparous species which guards its nest.

LEFT | The common name of the Ocellated Skink (*Chalcides ocellatus*) comes from the black-edged white spots on its back that look like tiny eyes.

HABITATS
Desert, semidesert, grassland, succulent veld, woodland, tropical forest, and plantations

SIZE
SVL 1¼ in (29 mm) Majunga Skink (*Madascincus nanus*) to 13⅓ in (340 mm) Western Forest Legless Skink (*Feylinia currori*)

ACTIVITY
Fossorial, semi-fossorial, or terrestrial; many species are diurnal but activity cycles are not known for all taxa

REPRODUCTION
Oviparous and viviparous genera, and genera with oviparous and viviparous species (*Plestiodon*). Arnold's Skink (*Proscelotes arnoldi*) can adopt either strategy. Strategy not known for many genera

DIET
Presumably small arthropods

SCINCIDAE—EGERNIINAE
SOCIAL SKINKS

The subfamily Egerniinae is confined to Australia, New Guinea, and the Solomon Islands, and consists of eight genera and 62 species. They are sometimes known as the social skinks, because some species are believed to be intelligent and able to recognize related individuals or participate in monogamous pair-bonding.

Included in the Egerniinae are some of the largest living skinks, such as the curiously named Land Mullet (*Bellatorias major*), the blue-tongued skinks (*Tiliqua*), which extend their broad, blue tongues as a defensive display, and the highly arboreal, primarily vegetarian, prehensile-tailed Monkey-tailed Skink (*Corucia zebrata*), from the Solomon Islands.

BELOW | Solomon Monkey-tailed Skinks (*Corucia zebrata*) are large bodied with prehensile tails. The females are good mothers but large individuals can be aggressive to conspecifics and may even kill smaller specimens.

EGERNIINAE

DISTRIBUTION
Australia, including Tasmania, New Guinea, and the Solomon Islands

GENERA
Bellatorias, Corucia, Cyclodomorphus, Egernia, Liopholis, Lissolepis, Tiliqua, and *Tribolonotus*

HABITATS
Rainforest, savanna woodland, gardens, scrubland, desert, semidesert, rocky escarpments, swamps, and plantations

SIZE
SVL 1½ in (40 mm) Blanchard's Crocodile Skink (*Tribolonotus blanchardi*) to 15⅓ in (391 mm) Land Mullet (*Bellatorias major*)

ACTIVITY
Fossorial, semi-fossorial, terrestrial, or arboreal; diurnal, crepuscular, or nocturnal

Also contained in this subfamily are several endemic Australian genera, the slender, elongate she-oak skinks (*Cyclodomorphus*), the desert-dwelling spiny skinks (*Egernia*) and desert skinks (*Liopholis*), and the swamp skinks (*Lissolepis*).

Probably the most unusual members of this subfamily are the crocodile skinks (*Tribolonotus*) of New Guinea and the Solomons. This genus is now known to comprise ten species, some of which resemble tiny dinosaurs. The Graceful Crocodile Skink (*T. gracilis*) of Papua New Guinea has a large, triangular head, with a bony crest at the back and four rows of raised, spiky scales down its back. The eye is surrounded by red pigment and includes a yellow iris. There are also unusual glands on the hands, feet, and abdomen, but their purpose is currently unknown. When disturbed, these small lizards shriek loudly, a sound belying their small size.

ABOVE | The Graceful Crocodile Skink (*Tribolonotus gracilis*) resembles a miniature dinosaur due to its angular head with its bony crest, facial pigmentation and four rows of keeled scales on the back.

While most social skinks are viviparous, at least seven species of *Tribolonotus* are oviparous, the females laying a single egg, which in the case of the Graceful Crocodile Skink may equal ten percent of the female's weight.

RIGHT | The threat display of the Eastern Blue-tongued Skink (*Tiliqua scincoides*) involves protruding its large blue-colored tongue while gaping its mouth widely at its perceived enemy.

REPRODUCTION
Most species are viviparous, producing one or rarely two neonates (*Corucia*), or up to 25 (*Cyclodomorphus*). *Tribolonotus* is primarily oviparous, the Graceful Crocodile Skink (*T. gracilis*) laying a single, large, striated, leathery-shelled egg; only one species, the Guadalcanal Crocodile Skink (*T. schmidti*) is known to be viviparous

DIET
Mostly arthropods, from termites to large insects, but *Tiliqua* is omnivorous, taking insects, snails, small vertebrates, eggs, fruit, and flowers, while *Corucia* is herbivorous, feeding on leaves and fruit; juveniles are coprophagic, eating the mother's feces in order to obtain the gut bacteria needed to process vegetation

SCINCIDAE—SPHENOMORPHINAE
AUSTRALO-ASIAN SKINKS

The Sphenomorphinae, including the Ristellidae, is the largest of the scincid subfamilies, with over 590 species in 36 genera. Most are Australasian or Asian, but one genus (*Scincella*) also occurs in Central America. The nominate genus (*Sphenomorphus*) contains 113 species of wedge and forest skinks, which are found from Southeast Asia to the Solomon Islands. Other large genera are centered on Australia, the striped skinks (*Ctenotus*, 102 species) and the sliders (*Lerista*, 99 species), which contains species with well-developed limbs, short, reduced limbs and elongate bodies, and entirely limbless fossorial species.

The subfamily also contains some unusual species. The Fojia Mountain Skink (*Fojia bumui*) from Papua New Guinea looks more like a long-legged anole lizard (*Anolis*, page 200) than a skink, and it also has areas of strange glandular tissue on the chin, femurs, abdomen, and tail (see also *Tribolonotus*, page 131). It inhabits rocky streams and is built for rock-hopping.

The five species in the genus *Prasinohaema* are the only reptiles to possess vivid green blood.

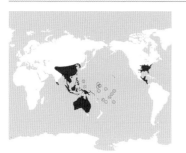

SPHENOMORPHINAE

DISTRIBUTION
Australia, New Guinea, Asia, and North and Central America

GENERA
Anomalopus, Asymblepharus, Calyptotis, Coeranoscincus, Coggeria, Concinnia, Ctenotus, Eremiascincus, Eulamprus, Fojia, Glaphyromorphus, Hemiergis, Insulasaurus, Isopachys, Kaestlea, Lankascincus, Larutia, Leptoseps, Lerista, Lipinia, Nangura, Notoscincus, Ophioscincus, Otosaurus, Papuascincus, Parvoscincus, Pinoyscincus, Prasinohaema, Ristella, Saiphos, Scincella, Silvascincus, Sphenomorphus, Tropidophorus, Tumbunascincus, and Tytthoscincus

HABITATS
Rainforest, sclerophyll forest, upland forest, desert, grassland, heathland, coastal dunes, riverine forest, swamps, and rocky piles

The coloration is a biliverdin-like pigment, as found in bile, which is believed to prevent the skinks from catching malaria because the parasites cannot survive in the blood. These skinks also have green mucous membranes, and prehensile tails.

LEFT | The Papuan Big-eyed Forest Skink (*Sphenomorphus annectans*) inhabits the damp mossy rainforest around Lake Kutubu in the Southern Highlands of Papua New Guinea.

ABOVE | An Eastern Striped Skink (*Ctenotus robustus*) perched on a vantage point. This lizard belongs to the second largest sphenomorphine genus (after *Sphenomorphus*).

Members of the Sphenomorphinae are fossorial, terrestrial, arboreal, and also aquatic. The rugose-skinned Asian water skinks (*Tropidophorus*) are found in close proximity to rivers and other watercourses. They closely resemble the Prickly Forest Skink (*Concinnia queenslandiae*) of Australia.

The Australian snake-tooth skinks (*Coeranoscincus*) possess long, recurved teeth that may assist them when eating earthworms. There are two species, one with four limbs, each with three toes, the Three-toed Snake-tooth Skink (*C. reticulatus*), and one completely limbless, the Limbless Snake-tooth Skink (*C. frontalis*).

SIZE
SVL 1 in (24 mm) Nicobar Ground Skink (*Scincella macrotis*) to 11½ in (290 mm) Limbless Snake-tooth Skink (*Coeranoscincus frontalis*)

ACTIVITY
Fossorial, semi-fossorial, terrestrial, arboreal, or aquatic (*Eulamprus, Tropidophorus*); diurnal, crepuscular, or nocturnal

REPRODUCTION
Oviparous and viviparous genera, and genera with both oviparous and viviparous species, including the Black-tailed Bar-lipped Skink (*Glaphyromorphus nigricaudis*) and Three-toed Skink (*Saiphos equalis*), which use both strategies. The reproductive strategy is unknown for many genera

DIET
Primarily arthropods, especially insects and spiders, but also earthworms (*Coeranoscincus*)

SCINCIDAE—MABUYINAE
SUN & GROUND SKINKS

Once the largest and most widely distributed genera in subfamily Mabuyinae, the nominate genus *Mabuya* contained over 150 species. These highly visible, primarily brown, terrestrial lizards were everywhere. But a revision of the genus split it into numerous smaller genera, the most important being *Trachylepis*, with over 87 species across Africa and the Indian Ocean (for example, the Seychelles Skink, *T. sechellensis*), and *Eutropis*, 44 species throughout Asia, including the Common Sun Skink (*E. multifasciata*), which is probably a species complex, occurring from India to Indonesia and introduced to Timor-Leste, New Guinea, and Florida.

Central American species were transferred to *Marisora*, and South American species to *Brasiliscincus* and *Varzea* and other genera, leaving only nine West Indian species in the formerly huge *Mabuya*, including the Greater Martinique Skink (*M. mabouya*).

Sun skinks can be extremely common; there may even be several species living in sympatry in areas of rich biodiversity, such as Borneo or the Amazon, and sometimes they are difficult to tell apart. As their common name suggests, these lizards love the sun. They inhabit open habitats such as grasslands or rocky outcrops, but also occur in rainforests, where they congregate along sunlit trails, in tree-fall sites, or where sunlight penetrates the canopy to illuminate the forest floor or a fallen tree trunk.

Most genera appear to be viviparous, but the arboreal Asian tree skinks (*Dasia*), one of the few genera never placed in *Mabuya*, are oviparous. *Trachylepis* and *Eutropis* contain both viviparous and oviparous species, and even species which can use either reproductive strategy.

BELOW | The Five-lined Skink (*Trachylepis quinquetaeniata*) is a common species across the northern half of Africa.

MABUYINAE

DISTRIBUTION
South America, West Indies, southeastern Europe, Africa, Madagascar, Asia, Cape Verde Islands, São Tomé and Príncipé, Socotran Archipelago, and other Indian Ocean islands

GENERA
Alinea, Aspronema, Brasiliscincus, Capitellum, Chioninia, Copeoglossum, Dasia, Eumecia, Eutropis, Exila, Heremites, Lubuya, Mabuya, Manciola, Maracaiba, Marisora, Notomabuya, Orosaura, Panopa, Psychosaura, Spondylurus, Toenayar, Trachylepis, Varzea, and Vietnascincus

HABITATS
Rainforest, woodland, grassland, rocky outcrops, swamps, and perianthropic habitats such as plantations and gardens

SIZE
SVL 1¼ in (32 mm) Ashwamedha Skink (*Eutropis ashwamedhi*) to 11¾ in (300 mm) Western Serpentiform Skink (*Eumecia anchietae*). The 13¾ in (350 mm) SVL Cape Verde Giant Skink (*Chioninia coctei*) is believed extinct

ACTIVITY
Terrestrial, arboreal, or fossorial; primarily diurnal

REPRODUCTION
Most genera are viviparous, but *Dasia* is oviparous and both *Eutropis* and *Trachylepis* contain viviparous and oviparous species, with several *Trachylepis* and *Eutropis* species using both strategies. The reproductive strategy is unknown for a few genera

DIET
Mostly arthropods, especially insects and spiders

ABOVE | The Olive Tree Skink (*Dasia olivacea*) is commonly seen basking on tree trunks across Asia from India to the Philippines and south to Indonesia.

SCINCIDAE—EUGONGYLINAE
AUSTRAL & SNAKE-EYED SKINKS

The subfamily Eugongylinae contains 454 species in 48 genera. One of the widest ranges is that of the Indo-Pacific snake-eyed skinks (*Cryptoblepharus*), from the African east coast, across the Indian Ocean, through the Indo-Australian Archipelago, and across the Pacific Ocean to Easter Island and the western coasts of Peru and Chile. There are more than 50 of these highly visible, small and slender, heliophilic skinks, which may be seen on rock faces, coconut palms, or the walls of buildings. They do not possess eyelids, but rather a snake-like spectacle over the eye. In Eurasia, from southeast Europe to Kazakhstan and Pakistan, they are replaced by the related Eurasian snake-eyed skinks (*Ablepharus*), and in Africa by the African snake-eyed skinks (*Panaspis*).

The majority of eugongyline genera are found across Indo-Australia and Oceania. The largest eugongyline genus is *Emoia* with 78 species, the most widely distributed species being the Mangrove Skink

LEFT | The Yule Island Skink (*Cryptoblepharus yulensis*) demonstates the snake-like, lidless, spectacle-covered eye that characterizes the snake-eyed skinks.

RIGHT | The Pacific Blue-tailed skink (*Emoia caeruleocauda*) continues a common theme among skinks and other lizards, such as some lacertids: the possession of an electric blue tail and stripes directing the attention of the predator to the tail, and away from the head. A skink can survive the loss of its tail and live to regenerate a new one.

EUGONGYLINAE

DISTRIBUTION
Australia including Tasmania, Norfolk and Lord Howe islands, New Guinea, Solomon Islands, New Caledonia, Vanuatu, New Zealand, and islands of the Indian and Pacific Oceans

GENERA
Ablepharus, Acritoscincus, Anepischetosia, Austroablepharus, Caesoris, Caledoniscincus, Carinascincus, Carlia, Celatiscincus, Cophoscincopus, Cryptoblepharus, Emoia, Epibator, Eroticoscincus, Eugongylus, Geomyersia, Geoscincus, Graciliscincus, Harrisoniascincus, Kanakysaurus, Kuniesaurus, Lacertaspis, Lacertoides, Lampropholis, Leiolopisma, Leptosiaphos, Liburnascincus, Lioscincus, Lobulia, Lygisaurus, Marmorosphax, Menetia, Morethia, Nannoscincus, Oligosoma, Panaspis, Phaeoscincus, Phasmasaurus, Phoboscincus, Proablepharus, Pseudemoia,

(*E. atrocostata*), which inhabits coastal mangrove swamps from New Guinea to the Japanese Ryukyu Islands, and the petite, electric-blue Pacific Blue-tailed Skink (*E. caeruleocauda*), which occurs widely across the Pacific. All New Zealand skinks (*Oligosoma*) and New Caledonian skinks (18 genera) belong to this subfamily. Several of the largest species, from New Caledonia and Mauritius, are believed extinct.

The Brown Sheen Skink (*Eugongylus rufescens*), from New Guinea, is a stout-bodied lizard of 6⅔ in (170 mm) SVL, which primarily feeds on large invertebrates, but will also take smaller skinks. One tiny species, the Dark-flecked Garden Skink (*Lampropholis delicata*) from New South Wales, has been accidentally introduced to New Zealand and Hawaii.

Pygmaeascincus, Saproscincus, Sigaloseps, Simiscincus, Tachygyia, Techmarscincus, and *Tropidoscincus*

HABITATS
Rainforest, grassland, rocky outcrops, freshwater swamps, mangroves (Mangrove Skink, *Emoia atrocostata*), and plantations

SIZE
SVL 1 in (25 mm) Wilson Bey's Snake-eyed skink (*Panaspis wilsoni*) to 8⅔ in (213 mm) White-banded Sheen Skink (*Eugongylus albofasciatus*). Both the larger 11¼ in (284 mm) Bocourt's Giant New Caledonian Skink (*Phoboscincus bocourti*) and 13⅔ in (340 mm) Mauritius Giant Skink (*Leiolopisma mauritana*) are believed extinct

ACTIVITY
Terrestrial, arboreal, or fossorial; primarily diurnal, many being basking sun-lovers, but others are secretive and nocturnal

REPRODUCTION
Most genera are oviparous, but those occurring in cooler southern latitudes (*Carinascincus, Pseudemoia*) or at higher elevations (*Lobulia*) are viviparous. Most species of the New Zealand genus *Oligosoma* are viviparous, but the Diving Skink (*O. suteri*) and the Lord Howe and Norfolk Islands Skink (*O. lichenigera*) lay eggs

DIET
Mostly arthropods, especially insects and spiders, but some larger species also take smaller skinks (*Eugongylus, Phoboscincus*)

SCINCIDAE—LYGOSOMINAE
AFRO-ASIAN SUPPLE, WRITHING & TREE SKINKS

ABOVE | Imagine being in a West African rainforest, sorting through piles of dark, wet leaf-litter, and then suddenly turning up this skink, the West African Fire-sided Skink (*Lepidothyris fernandi*). It is one of the most stunning of all skinks.

The Lygosominae is a relatively small subfamily containing only seven genera distributed through Africa and South and Southeast Asia. Most species are terrestrial or semi-fossorial and are referred to as either supple skinks (*Lygosoma, Riopa,* and *Subdoluseps*) or writhing skinks (*Mochlus*), but some of the relationships between and within genera are unclear. Island endemics include the Pemba Island Writhing Skink (*M. pembanus*), Mafia Island Writhing Skink (*M. mafianus*), and Bazaruto Island Writhing Skink (*M. lanceolatus*).

Many of the terrestrial species are small and fairly unicolor, with elongate bodies and short limbs, but the stocky and brightly colored fire-sided skinks (*Lepidothyris*) are larger and much more impressive. Some of the writhing skinks are also fairly large, the Tana River Writhing Skink (*M. mabuiiformis*) being the largest species in the subfamily (see below).

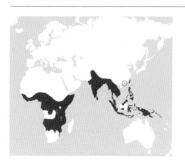

LYGOSOMINAE

DISTRIBUTION
Africa, and South and Southeast Asia

GENERA
Haackgreerius, Lamprolepis, Lepidothyris, Lygosoma, Mochlus, Riopa, and *Subdoluseps*

HABITATS
Rainforest, woodland, grassland, plantations, and semidesert

SIZE
SVL 1⅔ in (34 mm) Veun Sai Supple Skink (*Lygosoma veunsaiensis*) to 8¾ in (223 mm) Tana River Writhing Skink (*Mochlus mabuiiformis*)

ACTIVITY
Mostly terrestrial or semi-fossorial, but *Lamprolepis* is highly arboreal; primarily diurnal, and some species bask, but most are secretive

ABOVE | The highly arboreal, diurnal, Emerald Tree Skink (*Lamprolepis smaragdina*) occurs throughout Indonesia, Timor, and New Guinea.

In direct contrast to the terrestrial and burrowing species, the subfamily contains a single highly arboreal genus (*Lamprolepis*), which contains three species, although the stunning Emerald Tree Skink (*L. smaragdina*) is probably a species complex. Specimens from New Guinea are completely vivid green whereas specimens from Timor are either green anteriorly and gray-brown posteriorly, or entirely gray-brown.

Probably the most unusual member of the subfamily is the Haacke-Greer's Skink (*Haackgreerius miopus*). Known only from its holotype, this skink completely lacks forelimbs, though still possesses hind limbs. Its vestigial eyes are small and buried under undifferentiated ocular scales, suggesting it is poorly sighted and primarily fossorial in habit.

REPRODUCTION
Most species are oviparous, laying clutches of up to 12 parchment-shelled eggs, but *Mochlus* contains both oviparous and viviparous species, and the reproductive strategy of *Haackgreerius* is unknown

DIET
Arthropods, including insects and their larvae

LEFT | The Viviparous lizard (*Zootoca vivipara*) is one of the most widespread naturally occurring lizards in the world.

INFRAORDER LACERTOIDEA

The infraorder Lacertoidea contains two subgroups. The Teiioidea comprises four families: the Teiidae (teiids and tegus), Alopoglossidae (shade lizards), and Gymnophthalmidae (microteiids) all inhabit the Americas, while the Lacertidae (Old World lizards) dominate Europe, but are also present on the Canary Islands, and across Africa, Arabia, and Asia.

The second subgroup is the Amphisbaenia, the worm-lizards. They are usually treated as a third suborder, distinct from the Lacertilia (lizards) and the Serpentes (snakes). The Amphisbaenia are elongate and limbless (except *Bipes*), and split into six families. The largest family is the Amphisbaenidae from South America and Africa. The other five families are much smaller and geographically localized: the North African–Arabian Trogonophidae (four genera, six species); the Mediterranean Blanidae (a single genus, seven species); the Mexican Bipedidae, Cuban Cadeidae, and Florida Rhineuridae (single genera, with three, two, and one species, respectively).

TEIIDAE—CALLOPISTINAE & TEIINAE
AMEIVAS, WHIPTAILS & RACERUNNERS

The Callopistinae (Andean-Pacific teiids) contains two species of fast-moving, heliophilic lizards known as racerunners, which inhabit the arid Pacific coastal lowlands of South America from Ecuador to Chile, including the Atacama Desert. The small, yellow-spotted Inter-Andean Racerunner (*Callopistes flavipunctatus*) achieves a SVL almost twice that of the Chilean Spotted Racerunner (*C. maculatus*). It preys on mice, birds, and teiid lizards of genus *Dicrodon*. Resembling small Old World monitor lizards (*Varanus*, pages 220–231), they are sometimes called "false monitors." Some authors include *Callopistes* in the Tupinambinae.

The second teiid subfamily, Teiinae (ameivas, whiptails, and racerunners), contains over 154 species in 13 genera, from the southwestern United States to Argentina, and on the Caribbean islands. Many are highly visible, diurnal species that are seen basking or running in the sun in a wide variety of habitats, from deserts to rainforests. They rather resemble the Old World lacertid lizards (Lacertidae, pages 152–157) in their morphology and ecology.

The largest mainland species are the ameivas (*Ameiva*), but the most visible are the whiptails, such as the North American *Aspidoscelis*, and Central and South American *Cnemidophorus*. They are replaced in the Caribbean by the Caribbean racerunners (*Pholidoscelis*), a genus containing the largest and smallest members in the subfamily. Most species have smooth scales, but the genus *Kentropyx* contains the keel-scaled teiids, which possess weakly to strongly keeled scales. Many species show sexual dichromatism, some sexual dimorphism, and a few are all-female parthenogens.

While most South American taxa occur east of the Andes, this subfamily is represented west of the Andes by the sometimes herbivorous Pacific desert

CALLOPISTINAE

DISTRIBUTION (BLUE ON MAP)
Western Andean foothills, from Ecuador to central Chile

GENUS
Callopistes

HABITATS
Dry hilly woodland or savanna, semidesert, rocky coastlines, and ravines

SIZE
SVL 6¾in (173 mm) Chilean Spotted Racerunner (*Callopistes maculatus*) to 12¾ in (325 mm) Inter-Andean Racerunner (*C. flavipunctatus*)

ACTIVITY
Terrestrial; diurnal and heliophilic

REPRODUCTION
Both species are oviparous, producing up to six leathery-shelled eggs

DIET
Smaller reptiles, small mammals, birds, insects, crustaceans, and arachnids

LEFT | A large and brightly colored male Common Ameiva (*Ameiva ameiva*).

LEFT | The arid-habitat dwelling Inter-Andean racerunner (*Callopistes flavipunctatus*) resembles a small Old World monitor lizard in its appearance and diet.

RIGHT | The Western Tiger Whiptail (*Aspidoscelis tigris*) is represented by 15 subspecies across western USA and northern Mexico.

teiids (*Dicrodon*) and the Pacific Coastal Whiptail (*Medopheos edracantha*) of Peru and Ecuador. The only species demonstrating any form of limb reduction are the four-fingered teiids (*Teiius*) of southern South America.

TEIINAE

DISTRIBUTION (RED ON MAP)
North, Central, and South America

GENERA
Ameiva, Ameivula, Aspidoscelis, Aurivela, Cnemidophorus, Contomastix, Dicrodon, Glaucomastix, Holcosus, Kentropyx, Medopheos, Pholidoscelis, and *Teius*

HABITATS
Rainforest, woodland, Cerrado, desert, and open coastal habitats

SIZE
SVL 2 in (52 mm) Puerto Rican Blue-tailed Ameiva (*Pholidoscelis wetmorei*) to 9¾ in (250 mm) Dominican Ameiva (*P. fuscatus*)

ACTIVITY
Terrestrial and semi-arboreal (some *Kentropyx*); diurnal and heliophilic

REPRODUCTION
All species are oviparous, producing up to nine leathery-shelled eggs, several times a year. Some *Aspidoscelis* are parthenogenetic

DIET
Primarily arthropods, especially insects and spiders, but occasionally smaller lizards, and some juveniles eat cactus fruit (e.g. *Dicrodon*)

TEIIDAE—TUPINAMBINAE
TEGUS

The Tupinambinae contains four genera and 14 species. The genera *Tupinambis* and *Salvator* are the tegus, moderately large to large lizards that occupy the neotropical niches that in the Old World would be occupied by monitor lizards (*Varanus*, page 220–231). With their ruggedly built bodies, powerful legs, and large, pointed heads, the tegus can function as omnivores that occasionally feed on fruit, predators of smaller vertebrates, raiders of birds' nests, or scavengers of carrion. The 11 species are found in habitats ranging from desert to rainforest. Females may lay large clutches of eggs (up to 32) in burrows excavated in termite nests, which are then resealed by the termites. Incubation is relatively long for a squamate, at three to four months.

The Crocodile Tegu (*Crocodilurus amazonicus*) and the two caiman lizards (*Dracaena*) are arboreal and aquatic tegus that inhabit rivers, seasonally flooded

TUPINAMBINAE

DISTRIBUTION
South America

GENERA
Crocodilurus, Dracaena, Salvator, and *Tupinambis*

HABITATS
Rainforest, dry forest, Cerrado, lakes, rivers, flooded habitats, and desert

SIZE
SVL 12½ in (320 mm) Crocodile Tegu (*Crocodilurus amazonicus*) and Slender Tegu (*Tupinambis longilineus*) to 24¼ in (614 mm) Red Tegu (*Salvator rufescens*)

ACTIVITY
Terrestrial, aquatic, and arboreal (*Dracaena*); diurnal and heliophilic

REPRODUCTION
All species are oviparous, producing up to 32 leathery-shelled eggs

apó forests, swamps, lagoons, and lakes. The Crocodile Tegu is the smaller species (SVL ? in/320 mm), feeding on arthropods and smaller rtebrates, including fish. The larger caiman ards, the Guianan Caiman Lizard (*D. guianensis*, VL 16¼ in/412 mm) and Paraguayan Caiman zard (*D. paraguayensis*, SVL 17¾ in/450 mm), are en more aquatic, but they are specialist feeders on ollusks, which are crushed between powerful olars in their massively built jaws. The Guianan aiman Lizard is the better known of the two ecies; it produces small clutches of two to four gs. The Paraguayan Caiman Lizard has been very tle studied. The Guianan Caiman Lizard is a unning green, armored lizard.

DIET
Other reptiles, amphibians, small mammals, birds and their eggs, carrion (*Tupinambis*, *Salvator*), snails (*Dracaena*), insects, spiders, worms, and fish (*Crocodilurus*); occasional vegetation

LEFT | The Guianan Caiman Lizard (*Dracaena guianensis*) is a green lizard with a robust orange head. The raised scales down its back resemble the osteoderms of the South American alligatorines known as caiman.

TOP | An adult Common Tegu (*Tupinambis teguixin*) on the prowl with its snake-like forked tongue. Note the parasitic tick (Ixodida) on the side of the head.

ABOVE | The semi-aquatic Crocodile Tegu (*Crocodilurus amazonicus*) is one of the smallest tegus. Not to be confused with the unrelated Crocodile Lizard (*Shinisaurus crocodilurus*, page 218) from China.

ALOPOGLOSSIDAE
SHADE LIZARDS

BELOW | The Northern Shade Teiid (*Alopoglossus angulatus*) exhibits oblique rows of strongly keeled scales.

Until recently the Alopoglossidae comprised two genera (*Alopoglossus* and *Ptychoglossus*), but all species are now placed in the nominate genus *Alopoglossus*. These lizards were also contained within the Gymnophthalmidae, but they are more closely related to the Teiidae and are therefore now treated as a separate family, a sister taxon to the Teiidae. This is an unusual scenario given their contrasting sizes, with members of the Alopoglossidae having SVLs of 1⅓–3¼ in (33–80 mm), while the teiids range from 12 to 24¼ in (52–614 mm).

Shade teiids are secretive, diurnal inhabitants of leaf litter, but they can be distinguished by the condition and arrangement of their dorsal scales. They may possess smooth or weakly keeled scales arranged in transverse rows, or strongly keeled scales arranged in oblique rows, so may be referred to as smooth or keeled shade teiids. The genus *Alopoglossus* contains 28 species.

DISTRIBUTION
Costa Rica to the Brazilian and Peruvian Amazon and the Guianas

GENUS
Alopoglossus

HABITATS
Primary and secondary rainforest

SIZE
SVL 1¼ in (33 mm) Lehmann's Shade Lizard (*Alopoglossus lehmanni*) to 3¼ in (80 mm) Drab Shade Lizard (*A. copii*)

ACTIVITY
Terrestrial and semi-fossorial; diurnal and heliophobic

REPRODUCTION
All species are oviparous, producing a clutch of 1–2 leathery-shelled eggs

DIET
Small leaf-litter arthropods, such as insects and spiders

Shade teiids occur from Costa Rica, in Central America, southwards into South America as far as Peru and the Amazon in the west and the Guianas in the east. They inhabit rainforest leaf litter or low vegetation, especially in close proximity to streams and creeks, and despite being diurnal in habit they are not heliophiles, preferring shade to sunlight. They may also be active on moonlit nights. Being small lizards, they produce small clutches of eggs and feed on small leaf-litter invertebrates, while they are themselves the prey of larger lizards and snakes.

ABOVE | The Short-nosed Shade Teiid (*Alopoglossus brevifrontalis*) exhibits transverse rows of smooth scales. It also has frontal and frontonasal scales that are broader than they are long, hence its scientific name *brevifrontalis*.

GYMNOPHTHALMIDAE—GYMNOPHTHALMINAE, RIOLAMINAE & RHACHISAURINAE
SAND, FOREST, SPECTACLED & TEPUI MICROTEIIDS

The lizards in the Gymnophthalmidae are usually referred to as microteiids because they were once included in the Teiidae, which is populated by significantly larger lizards. The type genus of the Gymnophthalmidae is *Gymnophthalmus*, the spectacled microteiids, from *gymn*, meaning "naked," and *ophthalmus*, meaning "eyes," a reference to the condition of their eyes, which lack eyelids and are covered by transparent, snake-like spectacles. As it is currently recognized, the Gymnophthalmidae contains four subfamilies (see also Alopoglossidae, pages 146–7).

The Gymnophthalminae (sand, forest, and spectacled microteiids) contains 42 species in 18 genera. The genus *Gymnophthalmus* contains eight species of elongate, glossy, spectacled microteiids, which are among the most frequently encountered leaf-litter lizards in northern South America.

The subfamily also contains the specialized microteiids from the extensive sand dunes of the

LEFT | The Central American Spectacled Teiid (*Gymnophthalmus speciosus*) is the only Middle American member of its genus. Like snakes, they have lidless eyes covered by spectacles or brilles.

RIGHT | The Murisopán-tepui Teiid (*Riolama inopinata*) is endemic to that straight-sided Venezuelan tepui.

GYMNOPHTHALMINAE
DISTRIBUTION
South America

GENERA
Acratosaura, Alexandresaurus, Calyptommatus, Caparaonia, Colobodactylus, Colobosaura, Gymnophthalmus, Heterodactylus, Iphisa, Micrablepharus, Nothobachia, Procellosaurinus, Psilops, Rondonops, *Scriptosaura, Stenolepis, Tretioscincus,* and *Vanzosaura*

HABITATS
Lowland and montane rainforest, dry forest, savanna, Chaco, Cerrado, and riverine sand dunes

SIZE
SVL 1 in (27.5 mm) Four-fingered Sand Microteiid (*Procellosaurinus tetradactylus*) to 4⅓ in (110 mm) Rio de Janeiro Litter Microteiid (*Heterodactylus imbricatus*)

São Francisco River system in northeastern Brazil. These are the sand-dwelling microteiids of the genera *Calyptommatus*, *Procellosaurinus*, and *Psilops*. The sand-swimming microteiids of genus *Calyptommatus* are extremely elongate, snake-like, and lack limbs, while those in the other two genera look like more elongate *Gymnophthalmus*. Some species have bright red tails, which may be to distract the attention of predators away from the lizard's head and body. The São Francisco River system is an extremely interesting environment because it has a large and endemic psammophile (sand-dwelling) herpetofauna, including a number of unique snake and worm-lizard species.

The Riolaminae (Venezuelan tepui microteiids) and Rhachisaurinae contain six and one species, respectively, of small, leaf litter-dwelling tepui or montane microteiids. *Riolama* inhabit the Venezuelan-Brazilian, straight-sided mountains known as tepuis, while the Serra do Cipó Microteiid (*Rhachisaurus brachylepis*) occurs in Serra do Cipó, in Minas Gerais, eastern Brazil.

ACTIVITY
Terrestrial, semi-fossorial, and fossorial; diurnal

REPRODUCTION
All species are believed to be oviparous, producing a clutch of 1–2 leathery-shelled eggs

DIET
Small leaf-litter arthropods, such as insects and spiders

RIOLAMINAE & RHACHISAURINAE

DISTRIBUTION
Amazonas, Venezuela and Brazil (blue dots) and Minas Gerais, Brazil (yellow dot)

GENERA
Riolama and *Rhachisaurus*

HABITATS
Montane and tepui vegetation

SIZE
SVL: 1⅓ in (43 mm) Murisipán-tepui Microteiid (*Riolama inopinata*) to 2⅓ in (59 mm) Bright-bellied Tepui Microteiid (*R. luridventris*); 3½ in (88 mm) Large Tepui Teiid (*Riolama grandis*)

ACTIVITY
Terrestrial and semi-fossorial; diurnal

REPRODUCTION
Presumed oviparous

DIET
Presumably small arthropods

GYMNOPHTHALMIDAE—CERCOSAURINAE
SUN, SPINY & STREAM MICROTEIIDS

LEFT | The Kartabo Bachia (*Bachia flavescens*) from Guyana, has such an elongate body and short legs that its preferred form of locomotion is flick-leaping, forming a figure-of-eight and launching itself forward by a sudden uncoiling, repeating the same action repeatedly.

The Cercosaurinae contains over 212 species in 34 genera, and is named for the eyed microteiids (*Cercosaura*), which have eye-like markings along their flanks. *Cercosaura* is a large genus, with 16 species distributed from northern South America south to Uruguay and northern Argentina. In males of the related luminous microteiids (*Proctoporus*, *Petracola*, and *Riama*), the whites of these spots are said to glow in the dark. They are sometimes referred to as "lightbulb lizards."

The bachias (*Bachia*) are an unusual group of microteiids. These elongate lizards have extremely short limbs and move over the ground by "flick-leaping," forming the body and tail into a figure-eight shape, and then flicking themselves forward. Similar locomotion may be seen in the completely unrelated Australian flick-leapers (*Delma*, page 117), in the gekkotan family Pygopodidae. Other species, such as the water microteiids (*Neusticurus*), are found living in the leaf litter along jungle streams.

New species are being discovered in the leaf litter of South America with some frequency. In 2019 a new genus and species of microteiid was described from the eastern Andes of Peru, at around 9,190 ft (2,800 m). *Dendrosauridion yanesha*, which has no common name yet but might be called the Yanesha Tree Microteiid (the Yanesha are the indigenous

CERCOSAURINAE

DISTRIBUTION
Costa Rica to Ecuador west of the Andes, and Uruguay east of the Andes

GENERA
Adercosaurus, Amapasaurus, Anadia, Andinosaura, Anotosaura, Arthrosaura, Bachia, Centrosaura, Cercosaura, Colobosauroides, Dendrosauridion, Dryadosaura, Echinosaura, Ecpleopus, Euspondylus, Gelanesaurus, Kaieteurosaurus, Leposoma, Loxopholis, Macropholidus, Marinussaurus, Neusticurus, Oreosaurus, Pantepuisaurus, Petracola, Pholidobolus, Placosoma, Potamites, Proctoporus, Rheosaurus, Riama, Selvasaura, Wilsonosaura, and Yanomamia

HABITATS
Lowland and montane rainforest, trails and clearings, dry forest, rocky outcrops, and stream and swamp edges

SIZE
SVL 1¼ in (30 mm) Amapa Four-fingered Microteiid (*Amapasaurus tetradactylus*) to

ABOVE | Males of the Trinidad Lightbulb Lizard (*Riama shrevei*) exhibit black-edged white spots along their flanks that are said to glow in the dark, leading to the names luminous or lightbulb lizards.

LEFT | Eigenmann's Eyed Teiid (*Cercosaura eigenmanni*) inhabits the rainforest leaf-litter across Amazonian Brazil, Peru, and Bolivia.

4¾ in (121 mm) Colombian Water Microteiid (*Neusticurus medemi*)

ACTIVITY
Terrestrial, aquatic, semi-fossorial, and fossorial; diurnal, or occasionally nocturnal

REPRODUCTION
All species are believed to be oviparous, producing a clutch of 1–2 leathery-shelled eggs

DIET
Insects and spiders, some specialization on ants; also smaller lizards

people of this part of Peru), is an elongate, arboreal, and rather secretive lizard, with a tail almost twice the length of its body. The body and tail are covered in rectangular scales that make it look like a miniature alligator lizard (Gerrhonotinae, page 210). Another new genus and species, the Rio Bravo Microteiid (*Selvasaura brava*), had already been described from the Peruvian Andes in 2018.

LACERTIDAE—GALLOTIINAE
CANARY ISLAND LIZARDS & SAND RACERS

The Gallotiinae is a small subfamily of the Lacertidae (Old World lizards) containing just two genera and 14 species distributed across northwest Africa and southwest Europe. The nominate genus is *Gallotia*, the endemic Canary Island lizards. All the main islands are home to two species, one small and one large, the latter including the three largest living lacertid lizards, from El Hierro (El Hierro Giant Lizard, *G. simonyi*; SVL 19¾ in/502 mm), La Palma (La Palma Giant Lizard, *G. auaritae*; SVL 17½ in/444 mm), and Gran Canaria (Gran Canaria Giant Lizard, *G. stehlini*; SVL 14½ in/370 mm).

The El Hierro and La Palma species are listed as Critically Endangered by the IUCN, while the Gran Canaria Giant Lizard may be threatened by the recently introduced California Kingsnake (*Lampropeltis californiae*), a specialist reptile predator. The Canary Islands lack native snakes, so the lizards may be vulnerable to predation and extinction.

GALLOTIINAE

DISTRIBUTION
Southwestern Europe, northwestern Africa, and the Canary Islands

GENERA
Gallotia and *Psammodromus*

HABITATS
Islands, desert, semidesert, open woodland, and other arid habitats

SIZE
SVL 1¾ in (44 mm) West Iberian Sand Racer (*Psammodromus occidentalis*) to 19¾ in (502 mm) El Hierro Giant Lizard (*Gallotia simonyi*)

ACTIVITY
Terrestrial, or occasionally semi-arboreal; diurnal and heliophilic

LEFT | The El Hierro Giant Lizard (*Gallotia simonyi*), the largest living member of the Lacertidae, is listed as Critically Endangered by the IUCN.

ABOVE | The Algerian Sand Racer (*Psammodromus algirus*), sometimes called the Large Psammodromus, is the largest and most widely distributed member of its genus.

REPRODUCTION
All species are oviparous, producing a clutch of up to 12 leathery-shelled eggs, sometimes 2–3 times a year

DIET
Insects, spiders, fruit, and vegetation; also carrion and gastropods (*Gallotia*)

The giant *Gallotia* may also have been exploited by humans since prehistory, but the largest species (*G. goliath*, SVL 3 ft/0.9 m, TTL 4–5 ft/1.25–1.5 m) is only known from fossils and went extinct before the arrival of humans.

The genus *Psammodromus* contains six species of sand racers. Occurring through northwest Africa and southwest Europe, the largest and most widely distributed species is the Algerian Sand Racer (*P. algirus*), from Morocco, Algeria, Tunisia, Gibraltar, Spain, Portugal, Andorra, southwest France, and the Italian island of Isola dei Conigli, near Lampedusa. It has also established on Mallorca in the Balearic Islands, following its introduction in the 1990s. The Algerian Sand Racer has a SVL of 3¼ in (80 mm), but a TTL of 12¼ in (310 mm), due to its exceedingly long tail. Other species, such as the Spanish Sand Racer (*P. hispanicus*), are half its size.

LACERTIDAE—LACERTINAE
EURASIAN LIZARDS

The subfamily Lacertinae was the large sister taxon to the Gallotinae. It was divided into two tribes, the Eurasian Lacertini and the Afro-Asian Eremiadini, but authors now seem to prefer elevating these tribes to subfamily level. The Lacertinae has a primarily Mediterranean distribution that spreads eastward into the Middle East and Central Asia, although there is one Far East Asian and Southeast Asian genus, *Takydromus*, the Oriental grass lizards. The 24 species of keel-scaled, long-tailed lizards are the only representatives of the Lacertinae throughout most of their range from Amur, Russia, and Japan, to Bangladesh and Indonesia.

Many of the European genera were formerly subgenera of *Lacerta*, which is now reduced to only ten species, including the highly variably patterned Sand Lizard (*L. agilis*), which, although rare and localized in its distribution in the UK, is found across a huge swathe of territory from Europe to

LACERTINAE

DISTRIBUTION
Europe, and southwestern, Central, Southeast, and Far East Asia

GENERA
Algyroides, Anatololacerta, Apathya, Archaeolacerta, Dalmatolacerta, Darevskia, Dinarolacerta, Hellenolacerta, Iberolacerta, Iranolacerta, Lacerta, Parvilacerta, Phoenicolacerta, Podarcis, *Scelarcis, Takydromus, Teira, Timon,* and *Zootoca*

HABITATS
Heathland, sand dunes, grassland, maquis, riverbanks, rocky outcrops and ruins, coastal islands, and rocky mountains

SIZE
SVL 1¾ in (45 mm) Pygmy Keeled Lizard (*Algyroides fitzingeri*) to 10¼ in (260 mm) European Eyed Lizard (*Timon lepidus*)

the Lake Baikal region of Central Asia. Although the range of the Sand Lizard is impressive, it is eclipsed by that of the Viviparous Lizard (*Zootoca vivipara*), which occurs across Europe and Asia, from Ireland in the west to Sakhalin Island in the east, and from the Arctic Circle of Scandinavia to northern Spain. Across most of its range this lizard is viviparous, the only live-bearing member of the Lacertinae, but southern French and northern Spanish populations are oviparous.

Among the most attractive species are the male green-bodied, blue-throated Western and Eastern Green Lizards (*L. bilineata* and *L. viridis* respectively) and the green, blue eye-spotted European Eyed Lizard (*Timon lepidus*), but the most ubiquitous European lizards are the numerous species of wall, rock, and ruin lizards (*Podarcis*).

LEFT | The Sand Lizard (*Lacerta agilis*) is sexually dichromatic: males are bright green while females, like the one pictured, are primarily brown with black and white ocelli markings.

ACTIVITY
Terrestrial, arboreal, saxicolous, or semi-fossorial; diurnal and heliophilic

REPRODUCTION
Most genera and species are believed to be oviparous, laying 1–4 or up to 23 (Eastern Green Lizard, *Lacerta viridis*) leathery-shelled eggs, with some species of *Darevskia* reproducing parthenogenetically. The most northerly distributed species (Viviparous Lizard, *Zootoca vivipara*) is viviparous, producing 1–11 neonates, although populations in the south of its range are oviparous, producing 1–13 leathery-shelled eggs

DIET
Invertebrates, including insects and spiders, although large species (*T. lepidus*) prey on smaller lizards

TOP | The Viviparous Lizard (*Zootoca vivipara*) is the northern-most and most widely distributed lizard in the world. It is also the only viviparous species in the Lacertinae.

ABOVE | Adult Eyed Lizards (*Timon lepidus*) are the largest European lizards with a body pattern of green and black, with blue ocelli markings on the flanks.

LACERTIDAE—EREMIADINAE
AFRO-ASIAN LIZARDS

The Eremiadinae is primarily an African subfamily, with 15 of the 22 genera confined to the continent. However, the most widely distributed genus, *Eremias*, is absent from Africa, being distributed from the northern coast of the Black Sea, across Central Asia to the Korean Peninsula. It contains 40 species of Eurasian racerunners, alert, fast-moving, terrestrial lizards of arid habitats. The largest genus, with 44 species,

BELOW | Busack's Fringe-toed Lizard (*Acanthodactylus busacki*) is an arid-habitat lizard occurring in Morocco and Western Sahara.

ACONTINAE

DISTRIBUTION
Africa, Iberia, Arabia, and Western, South, Central, and East Asia

GENERA
Acanthodactylus, Adolfus, Atlantolacerta, Australolacerta, Congolacerta, Eremias, Gastropholis, Heliobolus, Holaspis, Ichnotropis, Latastia, Meroles, Mesalina, Nucras, Omanosaura, Ophisops, Pedioplanis, Philochortus, Poromera, Pseuderemias, Tropidosaura, and Vhembelacerta

HABITATS
Desert, semidesert, grassland, steppe, woodland, sand dunes, tropical forest, mountains, rocky outcrops, and coastal islands

SIZE
SVL 1⅔ in (35 mm) Elba Snake-eyed Lizard (*Ophisops elbaensis*) to 4¾ in (120 mm) Delalande's Sandveld Lizard (*Nucras lalandii*)

ABOVE | The Western Blue-tailed Gliding Lizard (*Holaspis guentheri*) is a brightly colored arboreal lizard that glides to safety when it feels threatened.

are the fringe-toed lizards (*Acanthodactylus*), occurring from North Africa to the Middle East and northwest India. They have fringes of scales along their toes to aid in running across loose sand dunes.

The Steppe Runner (*E. arguta*) occurs in Ukraine and Crimea, while the European Fringe-toed Lizard (*A. erythrurus*) inhabits Spain and Portugal. There is also a third European representative of this subfamily, the Elegant Snake-eyed Lizard (*Ophisops elegans*), found in Turkey and Greece, which possesses snake-like spectacles over the eyes, in lieu of eyelids.

The sub-Saharan eremiadines contain some unusual species, including two species of blue-tailed gliding lizards (*Holaspis*). These unique and highly specialized lizards, which exhibit brilliant blue spots down their tails, can expand their ribs and leap into the air, gliding from branch to branch, or tree to ground, their landing being cushioned by the fusion of joints in the third and fourth fingers. Probably the most stunning African species is the vivid green, arboreal Green Keel-bellied Lizard (*Gastropholis prasina*), from coastal Kenya and Tanzania.

Most Eremiadinae are oviparous, only *Eremias* containing any viviparous species, and some, such as the Common Rough-scaled Lizard (*Meroles squamulosus*), are extremely short-lived, hatching, breeding, and dying inside a single year.

ACTIVITY
Terrestrial, arboreal, saxicolous, or semi-fossorial; diurnal and heliophilic

REPRODUCTION
Most genera and species are oviparous, producing leathery-shelled eggs, but *Eremias* also contains at least seven viviparous species

DIET
Invertebrates, including insects and spiders, although large species (*Gastropholis*) prey on smaller lizards

AMPHISBAENIDAE
WORM-LIZARDS

The Amphisbaenia is a clade of elongate, limbless, short-tailed reptiles, with vestigial eyes and body scales arranged in annular rings, superficially resembling earthworms or snakes. They evolved from within the infraorder Lacertoidea, as the snakes did from the infraorder Anguimorpha. The Amphisbaenia is usually treated as a separate suborder, distinct from Lacertilia (lizards) and Serpentes (snakes). Whereas most elongate squamates have a reduced left lung, the amphisbaenians possess a reduced right lung. Amphisbaenidae is the largest of the six families, with 12 genera and 182 species.

The Amphisbaenidae has a wide distribution incorporating South America, east of the Andes, the Greater Antilles, and sub-Saharan Africa. The largest genus, with 102 species, is *Amphisbaena*, occurring from Panama to Argentina, and on the islands of Trinidad, Cuba, and Hispaniola. The most widely distributed South American species are the Black and White Worm-lizard (*A. fuliginosa*) and the Red Worm-lizard (*A. alba*), which is also the largest living amphisbaenian. This latter species is often associated with leaf-cutter ant nests, where it feeds on the beetle larvae in the ant-nest middens. *Amphisbaena* have rounded heads, but two other South American genera possess more specialized

DISTRIBUTION
South America, Greater Antilles, and sub-Saharan Africa

GENERA
Amphisbaena, Ancylocranium, Baikia, Chirindia, Cynisca, Dalophia, Geocalamus, Leposternon, Loveridgea, Mesobaena, Monopeltis, and *Zygaspis*

HABITATS
Rainforest, dry forest, moist savanna, and semi-desert

SIZE
SVL 4¼ in (105 mm) Sierra de Pitomba Worm-lizard (*Amphisbaena talisiae*) to 32 in (810 mm) Red Worm-lizard (*A. alba*)

ACTIVITY
Fossorial; diurnal

REPRODUCTION
All are oviparous, laying a clutch of 1–4 (*Chirindia*) elongate, leathery-shelled eggs, except the Liwale Round-snouted Worm-lizard (*Loveridgea ionidesii*) and

heads for digging, the keel-headed *Mesobaena* and the shovel-snouted *Leposternon*.

The other nine genera occur in Africa: *Ancylocranium*, *Chirindia*, *Geocalamus*, and *Loveridgea* in East Africa; *Baikia* and *Cynisca* in West Africa; and *Dalophia*, *Monopeltis*, and *Zygaspis* in southern Africa. They represent an assortment of round-headed, shovel-headed, keel-headed, sharp-snouted, and wedge-snouted species, presumably adapted for subterranean life in a variety of different substrates, ranging from wet soil to loose sand or hardened mud.

Worm-lizards have an evolutionary history extending back at least 65 million years, and the ancestral African taxa are believed to have reached the continent by rafting (crossing the ocean on floating debris) from the Americas 60 million years ago, when the North Atlantic was much narrower than it is today.

ABOVE | The Mpumalanga Round-headed Worm Lizard (*Zygaspis vandami*) is one of the African species whose ancestors are believed to have originated from the Americas.

BELOW | The Black and White Worm-lizard (*Amphisbaena fuliginosa*) is a widespread, common, and very characteristic species from South America.

LEFT | The White (or Red) Worm-lizard (*Amphisbaena alba*) is the largest known worm-lizard in the world. In Brazil it is called "cobra com duas cabeças"—snake with two heads, the other head being the blunt tail. The real head is pointed with two eyes.

Cape Wedge-snouted Worm-lizard (*Monopeltis capensis*) from Africa, which are viviparous, with litters of 2–3 neonates

DIET
Arthropods, especially termites or ants, although the larger species take beetles, centipedes, and also earthworms, and *A. alba* also preys on beetle larvae in leaf-cutter ant middens

BLANIDAE
MEDITERRANEAN WORM-LIZARDS

LEFT | The Iberian Worm-lizard (*Blanus cinereus*) is easily found under stones on the Iberian Peninsula.

RIGHT | The Five-fingered Ajolote (*Bipes biporus*) is one of the strangest looking of all squamate reptiles.

The Blanidae comprises the single genus *Blanus*, containing seven species of worm-lizards with their distributions centered on the Mediterranean. The Iberian Worm-lizard (*B. cinereus*) and Vandelli's Worm-lizard (*B. vandellii*) occur in Spain, Portugal, and Gibraltar; the Mettetal's Worm-lizard (*B. mettetali*) and Tingitana Worm-lizard (*B. tingitanus*) occur in Morocco, and the Mardin Worm-lizard (*B. alexandri*), Anatolian Worm-lizard (*B. strauchi*), and Eastern Worm-lizard (*B. aporus*) occur in Turkey, Syria, Lebanon, Iraq, and on a few Greek islands.

Mediterranean worm-lizards are common in grassland or semi-arable habitats, where they may easily be found sheltering under stones, except during hot weather when they may estivate at greater depth. In cool or wet weather they may venture onto the surface at night. They feed entirely on small invertebrates such as ants, termites, insect larvae, or spiders. A defensive posture adopted when they are picked up involves curling the body tightly around a person's finger, and they may even attempt to bite but they are, of course, completely harmless.

DISTRIBUTION
Southwestern Europe, northwestern Africa, and Western Asia

GENUS
Blanus

HABITATS
Woodlands, stony steppe, or arable habitats with moist soils, under stones

SIZE
SVL 7¾ in (200 mm) Mardin Worm-lizard (*Blanus alexandri*) to 10 in (254 mm) Iberian Worm-lizard (*B. cinereus*)

ACTIVITY
Fossorial; nocturnal

REPRODUCTION
Oviparous, with clutches of 1–3 leathery-shelled eggs

DIET
Small arthropods, including insects and spiders

BIPEDIDAE
AJOLOTES

Also known as mole-lizards, the three species in genus *Bipes* are among the strangest squamates known. They are elongate and short-tailed, with annular rings of scales, rounded heads, and vestigial eyes, like other amphisbaenians, but it is the presence of a pair of powerful forelimbs, combined with the complete lack of hind limbs, that provides them with the "mole" epithet and makes them stand apart; the presence of forelimbs but no hind limbs is extremely rare in reptiles and amphibians. Equipped with three, four, or five claws, depending on species, the forelimbs enable the worm-lizard to burrow rapidly through loose sand.

The Five-fingered Ajolote (*B. biporus*) is the best known of the three species. It occurs in southern Baja California, Mexico, living in self-constructed burrows in the substrate, with openings under mesquite bushes, other vegetation, and stones. Ajolotes are much feared by locals, who believe they will burrow into a human anus. The other two species occur in Guerrero and Michoacán, western Mexico.

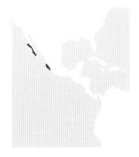

DISTRIBUTION
Northwestern Mexico

GENUS
Bipes

HABITATS
Desert, semidesert, and arid scrubland

SIZE
SVL 6⅔ in (160 mm) Three-fingered Ajolote (*Bipes tridactylus*) to 9½ in (240 mm) Five-fingered Ajolote (*B. biporus*) or Four-fingered Ajolote (*B. canaliculatus*)

ACTIVITY
Fossorial; diurnal or crepuscular

REPRODUCTION
Oviparous, with clutches of 1–4 leathery-shelled eggs

DIET
Arthropods

CADEIDAE
CUBAN WORM-LIZARDS

Two species of worm-lizards occur in Cuba. The Cuban Spotted Worm-lizard (*Cadea blanoides*) occurs in western Cuba and on the satellite Isla de la Juventud, an 850 sq mile (2,200 sq km) island south of the western Cuban mainland, with a questionable record from eastern Cuba, while the Cuban Sharp-nosed Worm-lizard (*C. palirostrata*) is endemic to Isla de la Juventud.

These worm-lizards are poorly studied. What little is known is that they prefer moderately warm conditions and the moist soil found under stones, planks, and fresh manure piles. They are also found in agricultural areas, where they are occasionally turned up by plowing. Both Cuban worm-lizards are believed to feed on small invertebrates. The Cuban Spotted Worm-lizard has many fewer rings of scales on its body (175–218) compared with the Cuban Sharp-nosed Worm-lizard (274–302), so the two are easily separated. However, the former may represent more than one species, as not all Cuban mainland specimens match its description.

RIGHT | The Cuban Spotted Worm-lizard (*Cadea blanoides*) is the most widely distributed member of this extremely poorly documented Cuban genus and family. This female has laid two elongate eggs.

DISTRIBUTION
Cuba, and Isla de la Juventud

GENUS
Cadea

HABITATS
Moist soils, including agricultural and rocky habitats

SIZE
SVL 9⅔ in (246 mm) Cuban Spotted Worm-lizard (*Cadea blanoides*) to 11 in (280 mm) Cuban Sharp-nosed Worm-lizard (*C. palirostrata*)

ACTIVITY
Fossorial; daily activity not known

REPRODUCTION
Oviparous, with clutches of 1–2 leathery-shelled eggs

DIET
Small invertebrates, probably termites and ants

RHINEURIDAE
FLORIDA WORM-LIZARD

A single species of worm-lizard, the Florida Worm-lizard (*Rhineura floridana*), inhabits central and northern Florida, from where it just enters Georgia. Pinkish in color, it has a flattened, shovel-shaped head and a short, flattened tail. The genetic structure and morphology of Florida Worm-lizards across their entire range suggest that they may actually represent a species complex of four to five species.

BELOW | The Florida Worm-lizard (*Rhineura floridana*) is the only amphisbaenian to occur within the United States.

The Florida Worm-lizard's preferred habitats include pine or turkey oak scrub ridges on sandy soils. It also occurs in hardwood hammocks, upland scrub, and fallow fields, provided they are well drained, and it is rarely seen on the surface unless its burrow has been flooded, which is when it is most susceptible to predation. Most specimens are found during excavations of the soil. Florida Worm-lizards feed on earthworms, termites, ants, and large, burrow-dwelling wolf spiders.

Florida Worm-lizards are vulnerable to flooding, habitat destruction, and attack by aggressive, invasive Fire Ants (*Solenopsis invicta*).

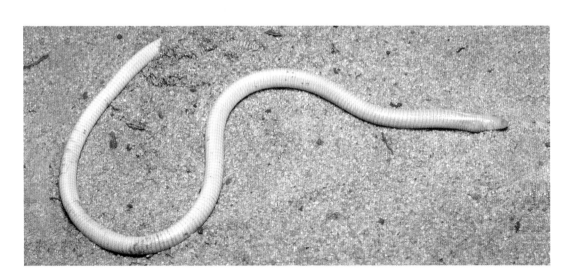

DISTRIBUTION Central and northern Florida, USA	**SIZE** SVL 15 in (380 mm) Florida Worm lizard (*Rhineura floridana*)
GENUS *Rhineura*	**ACTIVITY** Fossorial; daily activity not known
HABITATS Dry, sandy upland hammocks of pine or turkey oak, hardwood hammocks, upland scrub, and fallow fields	

REPRODUCTION
Oviparous, with clutches of 1–3 leathery-shelled eggs

DIET
Earthworms, termites, ants, and wolf spiders

TROGONOPHIDAE
AFRO-ARABIAN WORM-LIZARDS

The Trogonophidae contains four genera and six species distributed across North Africa and the Arabian Peninsula. The genus *Agamodon* contains three species, two from Somalia and one from Yemen. They are stocky-bodied, with short, shovel-shaped heads and a mottled pattern on the dorsum.

BELOW | The Checkered Worm-lizard (*Trogonophis wiegmanni*) from northwest Africa, is the only viviparous species in the family.

DISTRIBUTION
Northwestern and eastern Africa, Arabia, and Socotra

GENERA
Agamodon, Diplometopon, Pachycalamus, and *Trogonophis*

HABITATS
Desert, woodland, rocky wadis, cultivated land, and mountains with moist or sandy soils

SIZE
SVL 4 in (103 mm) Flat Worm-lizard (*Agamodon compressus*) to 9½ in (240 mm) Checkered Worm-lizard (*Trogonophis wiegmanni*)

ACTIVITY
Fossorial; nocturnal to crepuscular

BELOW | The Arabian Worm-lizard (*Diplometopon zarudnyi*) is a sandy desert dwelling species with an acutely flattened head for burrowing.

The other three genera are currently monotypic. However, the Checkered Worm-lizard (*Trogonophis wiegmanni*) of northwestern Africa, which is patterned with black and white or black and yellow, is divided into two subspecies, an eastern nominate form (*T. w. wiegmanni*) and a western form (*T. w. elegans*). Molecular studies support the morphological differences between these two populations, which may represent distinct species, with a third distinctive but unnamed population occurring on the Morocco–Algeria border.

REPRODUCTION
All genera and species are oviparous, laying leathery-shelled eggs, except *T. wiegmanni*, which is viviparous, with litters of 2–5 large (up to 3½ in/88 mm) neonates

DIET
Small invertebrates, and occasionally small geckos

The Arabian Worm-lizard (*Diplometopon zarudnyi*) is native to the eastern Arabian Peninsula, from Kuwait to Oman, and southwestern Iran. It too has a short, wedge-shaped head. The Socotra Worm-lizard (*Pachycalamus brevis*) is a little-known two-tone species, with a slightly more pointed snout, from the Socotran Archipelago.

The Checkered Worm-lizard is found under stones in woodland or agricultural habitats, in contrast to the Arabian Worm-lizard, which shelters under stones or pieces of tin sheeting on dry, sandy desert flats and on sand dunes. The Socotra Worm-lizard has been found under stones, rotten wood, and goat droppings. All known species appear to feed on invertebrates, such as crickets, but the Checkered and Arabian Worm-lizards also take small geckos.

LEFT | Unlike its mythical namesake, the Plumed Basilisk (*Basiliscus plumifrons*) cannot deal death with a single glance, but it can run across water.

INFRAORDER IGUANIA

The infraorder Iguania contains at least 14 families of toxicoferan lizards that can be split into two major groups, the Acrodonta, which have their teeth arranged along the top ridge of their jaws, similar to the arrangement in the unrelated Tuatara (*Sphenodon punctatus*, page 88), and the Pleurodonta, which have their teeth arranged along the inner sides of the jaws (see page 205).

The Acrodonta contains the Old World families Agamidae and Chamaeleonidae, from Africa, Europe, Asia, and Australasia. The Agamidae contains six subfamilies, but some authors elevate the Uromastycinae and Leiolepidinae to family status. No subfamilies of the Chamaeleonidae are currently recognized.

The remaining 12 families are contained in the Pleurodonta, and they were at one time treated as subfamilies of the Iguanidae. They are all New World lizards, from the Americas, except for the Fijian and Galapagos iguanas, the Opluridae of Madagascar, and some members of the Tropiduridae that are also found on the Galapagos Islands.

AGAMIDAE—AGAMINAE
AFRO-ASIAN AGAMAS

The subfamily Agaminae contains agamas that live in arid habitats, such as sandy or rocky desert, salt pans, dry woodland, savannas, and rocky outcrops and grassy areas surrounded by rainforest in West Africa. Some even live in human dwellings. The largest and most widely distributed African genus is *Agama*, a primarily terrestrial or saxicolous genus that occurs throughout the

BELOW | The Secret Toad-headed Agama (*Phrynocephalus mystaceus*), from Central Asia, has a defensive threat display that makes its mouth look much larger and as if it is lined with white teeth.

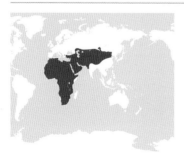

AGAMINAE

DISTRIBUTION
Africa, southeastern Europe, Arabia, and Asia

GENERA
Acanthocercus, Agama, Bufoniceps, Laudakia, Paralaudakia, Phrynocephalus, Pseudotrapelus, Stellagama, Trapelus, and Xenagama

HABITATS
Dry woodland, savanna, desert, semidesert, and mountains to 14,800 ft (4,500 m) on the Tibetan Plateau (Tibetan Toadhead Agama, *Phrynocephalus erythrurus*) or 16,400 ft (5,000 m) in Kashmir (Tuberculate Agama, *Laudakia tuberculata*)

SIZE
SVL 1⅔ in (42 mm) Rose-shouldered Toadhead Agama (*Phrynocephalus interscapularis*) to 8¼ in (210 mm) Rough-tailed Rock Agama (*Stellagama stellio*)

BELOW | A male Taylor's Shield-tailed Agama (*Xenagama taylori*), from Somalia and Ethiopia, with blue display markings on the lips and throat, and a spiny tail that can be useful for blocking its burrow.

continent and into Arabia. East and southern Africa are also inhabited by the arboreal tree agamas (*Acanthocercus*), while the four species of squat, terrestrial, shield-tailed agamas (*Xenagama*), which have short, flattened, spiny tails, are unique to Ethiopia and Somalia.

The ground agamas of genus *Trapelus* occur from Western Sahara, on the Atlantic coast of North Africa, to western China, while the related genus *Pseudotrapelus* inhabits northeast Africa and the Middle East.

The primarily Asian genera include the toadhead agamas (*Phrynocephalus*) and rock agamas (*Laudakia* and *Paralaudakia*). These three genera are concentrated in Central Asia but extend westward to Arabia. Several species occur at high elevations in the Himalayas or on the Tibetan Plateau. Only one monotypic genus enters extreme southeastern Europe, the Rough-tailed Rock Agama (*Stellagama stellio*), the largest species in the subfamily, which occurs in Greece, while the sole member of another monotypic genus, the Rajasthan Toad-headed Agama (*Bufoniceps laungwalaensis*), occurs along the Pakistan–India border.

Most Afro-Asian agamas are desert grays and browns, but the breeding males of some species are gaudily colored with red, orange, or blue heads and forelimbs. They seek to attract mates or defend territories with elaborate displays of head-bobbing or push-ups. All species feed on arthropods, many subsisting almost entirely on a diet of ants, an abundant food source in such arid environments.

ACTIVITY
Terrestrial, arboreal, and saxicolous; diurnal and heliophilic

REPRODUCTION
All genera are oviparous, except *Phyrnocephalus*, which contains both oviparous and viviparous species. Oviparous species lay oval, soft-shelled eggs

DIET
Arthropods, from ants to beetles, grasshoppers, millipedes, spiders, or scorpions

AGAMIDAE—AMPHIBOLURINAE
AUSTRALIAN & MELANESIAN DRAGONS

Australia is alive with agamid lizards belonging to the subfamily Amphibolurinae. Although called dragons, most are small species (*Ctenophorus* and *Diporiphora*) that guard territories, display to mates, or bask on rocks or atop bunches of tussock grass, some species being slender and striped to blend in with their grassy surroundings. The tree dragons (*Amphibolurus*) are found in Australia's dry woodlands, while the larger but elusive forest dragons (*Hypsilurus*) inhabit the rainforests of New Guinea, Queensland, eastern Indonesia, and the Solomon Islands.

One of the largest species is the Water Dragon (*Intellagama lesueurii*) of eastern Australia, which has a SVL of 9⅔ in (245 mm), large enough to feed on small vertebrates. For many years this semi-aquatic species was included in the genus *Physignathus* with the sole Southeast Asian member of the subfamily, the Thai Water Dragon (*P. cocincinus*), which inhabits Thailand, Vietnam, Cambodia, and southern China.

AMPHIBOLURINAE

DISTRIBUTION
Australia, New Guinea, Solomon Islands, and Southeast Asia

GENERA
Amphibolurus, Chelosania, Chlamydosaurus, Cryptagama, Ctenophorus, Diporiphora, Gowidon, Hypsilurus, Intellagama, Lophognathus, Lophosaurus, Moloch, Physignathus, Pogona, Rankinia, Tropicagama, and Tympanocryptis

HABITATS
Desert, semidesert, grassland, dry woodland, and tropical rainforest

SIZE
SVL 1⅓ in (34 mm) Crystal Creek Two-lined Dragon (*Diporiphora convergens*) to 10¼ in (258 mm) Frilled Lizard (*Chlamydosaurus kingii*)

LEFT | One of the most recognizable lizards in the world, the Frilled Lizard (*Chlamydosaurus kingii*) will spread the large frill around its neck and gape with its mouth when it feels threatened.

ABOVE | The Moloch (*Moloch horridus*) is a small, slow-moving, ant-eating dragon from the western deserts of Australia. The *horridus* part of its name means prickly, not horrid.

ACTIVITY
Terrestrial, arboreal, or semi-aquatic (*Intellagama, Physignathus*); diurnal and heliophilic

REPRODUCTION
All genera are oviparous, laying 2–30 parchment-shelled eggs

DIET
Arthropods, especially ants (*Moloch*); also fruit (*Pogona*) and small vertebrates (*Intellagama*)

The popular pet-trade bearded dragons (*Pogona*) are also members of this subfamily, but the most instantly recognizable species must be the Frilled Lizard (*Chlamydosaurus kingii*) of northern Australia and southern New Guinea. This species is well known for its startling ability to erect the large frill around its head to intimidate enemies; but if this strategy fails it will sprint bipedally, like a basilisk (*Basiliscus*, page 195), for the nearest tree, its frill folded away like a closed umbrella. Another distinctive dragon with an American counterpart is the Thorny Devil (*Moloch horridus*), a small, slow-moving, excessively spiny desert-dweller that feeds entirely on ants, and mirrors the American horned lizards (*Phrynosoma*, page 202).

AGAMIDAE—DRACONINAE
ASIAN DRAGONS

High levels of endemism exist within the South and Southeast Asian Draconinae, especially in the forests of southern and northeast India, Sri Lanka, Borneo, other Indonesian islands, and in the Philippines. Many species are poorly known in nature, and as habitat loss accelerates some may be lost even before they have been discovered and described for science.

The most commonly encountered species are the garden lizards (*Calotes*), also known as "bloodsuckers" because males often have red around the mouth. The most widely distributed and adaptable of these slender, diurnal lizards is the Common Garden Lizard (*C. versicolor*), occurring naturally from Iran to Indonesia (Sumatra), and introduced to Oman, Kenya, the Laccadives, Maldives, and Seychelles, Mauritius, and Florida.

Many species have unusual nasal adornments. For example, both sexes of the Sri Lankan Lyre-headed Lizard (*Lyriocephalus scutatus*) exhibit a round, ball-like protuberance, while male nose-horned lizards (*Harpesaurus*) from Indonesia possess a spiky nasal projection. The Sri Lankan horned lizards (*Ceratophora*) exhibit a range of species-specific nasal projections.

The most famous members of the subfamily are the 40 species of diminutive flying dragons (*Draco*) that occur from India to Timor. These tiny lizards do not actually fly; they glide to escape predators using expandable flaps of lateral skin, known as patagia, supported by extended ribs. When not in use, the patagium is closed like an umbrella, similar to the frill of the Frilled Lizard (*Chlamydosaurus kingii*, page 170).

LEFT | A Common Flying Dragon (*Draco volans*) gliding from a branch. These tiny lizards may travel almost 100 ft (30 m) in one glide.

RIGHT | Male Pondichéry Fan-throated Lizards (*Sitana ponticeriana*), from India, possess a large colorful dewlap which they use to display territorially.

DRACONINAE

DISTRIBUTION
South and Southeast Asia

GENERA
Acanthosaura, Aphaniotis, Bronchocela, Calotes, Ceratophora, Complicitus, Cophotis, Coryphophylax, Cristidorsa, Dendragama, Diploderma, Draco, Gonocephalus, Harpesaurus, Hypsicalotes, Hypsilurus, Japalura, Lophocalotes, Lyriocephalus, Malayodracon, Mantheyus, Microauris, Monilesaurus, Otocryptis, Phoxophrys, Psammophilus, Pseudocalotes, Pseudocophotis, Ptyctolaemus, Salea, Sarada, and *Sitana*

HABITATS
Tropical rainforest, dry tropical forest, rocky outcrops, gardens, and streams

SIZE
SVL 1½ in (39 mm) Suklaphantah Fan-throated Lizard (*Sitana schleichi*) to 7¼ in (185 mm) Sri Lankan Lyre-headed Dragon (*Lyriocephalus scutatus*)

ACTIVITY
Terrestrial, arboreal, or saxicolous; diurnal and heliophilic

REPRODUCTION
Most genera are oviparous, laying up to 20 parchment-shelled eggs, except *Cophotis*, which is viviparous, producing 2–8 neonates

DIET
Arthropods, especially ants or spiders; also earthworms and vegetation (*Lyriocephalus*)

Many dragons also possess an extendable flap under the throat known as a dewlap. Male fan-throated lizards (*Sitana*) display using colorful red, black, and blue dewlaps, often standing tall on their hind feet for maximum effect. This behavior mirrors that of the unrelated American anoles (*Anolis*, page 200).

AGAMIDAE—HYDROSAURINAE & LEIOLEPIDINAE
SAILFIN LIZARDS & BUTTERFLY LIZARDS

ABOVE | The Philippine Sailfin Lizard (*Hydrosaurus pustulatus*) is the smallest member of the Hydrosaurinae, but with its three crests it is still an impressive, primeval looking dragon.

The Hydrosaurinae is a monotypic subfamily comprising the genus *Hydrosaurus*, which contains five large species commonly known as sailfin lizards due to the presence of large, raised crests on the rear of the head, the back, and especially the tail. They comprise the Philippine Sailfin Lizard (*H. pustulatus*), Central Moluccan Sailfin Lizard (*H. amboinensis*), from the central Moluccas and New Guinea, Central Sulawesi Sailfin Lizard (*A. celebensis*), Southern Sulawesi Sailfin Lizard (*A. microlephus*), and Northern Moluccan Sailfin Lizard (*A. weberi*). The Philippine Sailfin Lizard is considered Vulnerable by the IUCN due to habitat loss and overcollection.

Sailfin lizards are semi-aquatic and arboreal, inhabiting riverine rainforest and mangroves, and basking on branches, ready to plunge into the water should danger threaten. They swim with ease, even in strong currents, using the laterally compressed tail, and can also run across the water for short distances in the manner of the neotropical basilisks (*Basiliscus*, page 195). They are primarily herbivorous, eating

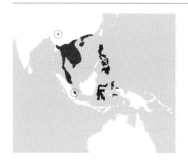

UROMASTYCINAE

DISTRIBUTION (BLUE ON MAP)
Southeast Asia

GENUS
Hydrosaurus

HABITATS
Riverine rainforest and mangrove forest

SIZE
SVL 11 in (280 mm) Philippine Sailfin Lizard (*Hydrosaurus pustulatus*) to 13¾ in (360 mm) Central Sulawesi Sailfin Lizard (*H. celebensis*)

ACTIVITY
Arboreal and semi-aquatic; diurnal

REPRODUCTION
All five species are oviparous, laying up to 12 eggs and multi-clutching in a single year

DIET
Vegetation, fruit (especially figs), and insects

IGUANIA—Agamas, chameleons, and iguanas

fruit, leaves, and seeds, with a preference for figs, but juveniles, and sometimes adults, also eat insects.

The Leiolepidinae is another small, monotypic agamid subfamily, although some authors treat it as a separate family. The nine species in genus *Leiolepis* are known as butterfly lizards because their brightly colored flanks are marked with black and yellow or orange, like the wings of a butterfly. These markings are exposed during displays of territoriality.

Butterfly lizards inhabit coastal or open, sandy lowland woodlands, from Myanmar to Sumatra. They are diurnal and terrestrial, spending much of their time in burrows, to which they will flee if danger threatens. They feed primarily on ground-growing plants, the Spotted Butterfly Lizard (*L. guttata*) demonstrating a preference for crocus flowers, but crabs and insects are also said to feature in their diets. Most species are sexual, but at least three, including the Thai Butterfly Lizard (*L. triploida*), are parthenogenetic.

BELOW | The Spotted Butterfly Lizard (*Leiolepis guttata*) uses its brightly marked flanks to display territorially and runs to its burrow if it feels threatened.

LEIOLEPIDINAE

DISTRIBUTION (RED ON MAP)
Southeast Asia

GENUS
Leiolepis

HABITATS
Coastal and lowland woodland

SIZE
SVL 5 in (126 mm) Böhme's Butterfly Lizard (*Leiolepis boehmei*) to 9¾ in (250 mm) Spotted Butterfly Lizard (*L. guttata*)

ACTIVITY
Terrestrial and fossorial; diurnal

REPRODUCTION
All species are oviparous, laying 1–8 eggs; some species are parthenogenetic

DIET
Vegetation, especially crocuses; also crabs and insects

AGAMIDAE—UROMASTYCINAE
DHAB LIZARDS & MASTIGURES

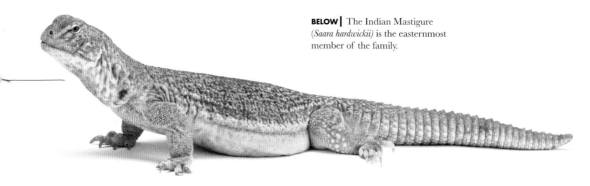

BELOW | The Indian Mastigure (*Saara hardwickii*) is the easternmost member of the family.

The Uromastycinae, treated as family Uromastycidae by some authors, is a small subfamily containing two genera and 18 species of relatively large, dorsoventrally compressed lizards that inhabit arid habitats across North Africa, Arabia, and South Asia. The Afro-Arabian species in genus *Uromastyx* are known as dhab lizards, while the Asian species in genus *Saara* may be called mastigures. All are stout-bodied lizards, with broad heads and exceedingly spiny tails. Individuals from the two genera can be distinguished by the presence (*Saara*) or absence (*Uromastyx*) of several rows of tiny scales between the spiny whorls of the tail. Tail lengths vary greatly between species, from the Horn of Africa Dhab Lizard (*U. princeps*), which possesses a short tail equivalent to 35–53 percent of its SVL, to the Saharan Dhab Lizard (*U. geyri*,) which has a tail of almost equal length to its SVL.

The North African Dhab Lizard (*U. acanthinura*) occurs along the northern edge of the Sahara and extends deep into the desert along wadis and mountain ranges. The Egyptian Dhab Lizard (*U. aegyptia*) ranges from Egypt into the Arabian Peninsula, as far southeast as UAE and Oman, and into Iraq, where its range overlaps that of the westernmost *Saara* species, the Mesopotamian Mastigure (*S. loricata*). The other Asian species are the Iranian Mastigure (*S. asmussi*), which inhabits Iran, Afghanistan, and Pakistan, and the Indian Mastigure (*S. hardwickii*), from Afghanistan to northwest India.

All dhab lizards and mastigures are primarily herbivorous, feeding on a wide range of ground-growing plants, but they also feed on insects. They are all oviparous.

UROMASTYCINAE

DISTRIBUTION
Africa, Arabia, and South Asia

GENERA
Saara and *Uromastyx*

HABITATS
Desert, semidesert, and rocky slopes

SIZE
SVL 4¾ in (122 mm) Macfadyen's Dhab Lizard (*Uromastyx macfadyeni*) to 16½ in (418 mm) Egyptian Dhab Lizard (*U. aegyptia*)

ACTIVITY
Terrestrial and fossorial; diurnal and heliophilic

REPRODUCTION
All species are oviparous, laying up to 23 eggs, once or twice a year

DIET
Vegetation and insects

LEFT | One of the most brightly colored dhab lizards is the North African Dhab Lizard (*Uromastyx acanthinura*). They may be red, orange, yellow, or green on the back with contrastingly dark head and tails.

BELOW | The Egyptian Dhab Lizard (*Uromastyx aegyptia*) is widely distributed from Egypt to Iran and Oman.

CHAMAELEONIDAE
EUROPEAN, ARABIAN & ASIAN CHAMELEONS

Chameleons are instantly recognizable because of their curious adaptations for life in the trees. Their bodies are laterally compressed and they possess zygodactylous feet, the five toes of each foot being fused into two and three toes, to provide balance when walking along fine twigs. They have bulging turret-eyes, with narrow eye openings, which function independently, enabling the chameleon to scan for prey or predators. They can then bring both eyes together to focus on a single point. Prey is approached stealthily using a swaying motion until the chameleon is close enough to propel its long, sticky tongue, as long as the chameleon's body, to capture it. They are also

DISTRIBUTION
Africa, Europe, Arabia, South Asia, Madagascar, and Indian Ocean

GENERA
Archaius, Bradypodion, Brookesia, Calumma, Chamaeleo, Furcifer, Kinyongia, Nadzikambia, Palleon, Rhampholeon, Rieppeleon, and *Trioceros*

HABITATS
Rainforest, dry woodland, thorn forest, karst limestone forest, and desert

SIZE
SVL ¾ in (20 mm) Nosy Hara Dwarf Chameleon (*Brookesia micra*) to 11⅔ in (295 mm) Parson's Chameleon (*Calumma parsonii*)

ACTIVITY
Arboreal or terrestrial; diurnal

LEFT | The Indian Chameleon (*Chamaeleo zeylanicus*) is the only truly Asian chameleon species, occurring across the Indian Peninsula and Sri Lanka.

ABOVE | A male Yemeni Veiled Chameleon (*Chamaeleo calyptratus*) uses its excellent eyesight to locate its prey before projecting its long sticky tongue over a distance equal to its own body length.

famous for their ability to undergo rapid color changes, for camouflage, or during display or male combat. Many species have long, prehensile tails that serve as additional limbs when climbing, and most also exhibit a diverse array of crests, horns, and casques which further enhance their prehistoric appearances.

Most of the 217 species of chameleon occur in Africa or Madagascar, but a few are found elsewhere.

REPRODUCTION
Most genera are oviparous; all South African *Bradypodion* are viviparous, while *Trioceros* contains both oviparous and viviparous species

DIET
Arthropods; larger species also prey on small vertebrates

The Veiled Chameleon (*Chamaeleo calyptratus*) inhabits Yemen and southwestern Saudi Arabia, while the Arabian Chameleon (*C. arabicus*) occurs in Yemen and Oman. In the Indian Ocean are found the Socotra Chameleon (*C. monachus*), Seychelles Chameleon (*Archaius tigris*), Grande Comore Chameleon (*Furcifer cephalolepis*), and Mayotte Chameleon (*F. polleni*). The Indian Chameleon (*C. zeylanicus*) inhabits India and Sri Lanka.

The Mediterranean Chameleon (*C. chamaeleon*) occurs in North Africa and in the Iberian Peninsula, Italy, Malta, Greece, and Cyprus, and the African Chameleon (*C. africanus*) is also found in Greece; both were introduced in ancient times. Jackson's Three-horned Chameleon (*Trioceros jacksonii*), from Africa, has been introduced to Hawaii, half a world away.

CHAMAELEONIDAE
AFRICAN CHAMELEONS

Chameleons are found throughout the African continent, excluding the Sahara Desert, from the Mediterranean to the Cape. They inhabit the dry forests of eastern and southern Africa, the rainforests of West Africa, riverine woodland, and gardens, with high levels of local endemism, especially on mountain ranges. Seven genera are represented in Africa.

One of the most recognizable chameleons is Jackson's Three-horned Chameleon (*Trioceros jacksonii*) of East Africa, but probably the most commonly encountered species is the widespread Flap-necked Chameleon (*Chamaeleo dilepis*), while the Cape Dwarf Chameleon (*Bradypodion pumilum*) may be found in the gardens and parks of the Cape, South Africa.

Pygmy chameleons of the genus *Rhampholeon*, cryptically patterned to resemble dead leaves, inhabit low rainforest vegetation from Cameroon to Uganda and Mozambique, but are replaced in the

LEFT | Pygmy chameleons like this Bearded Pygmy Chameleon (*Rieppeleon brevicaudatus*) is cryptically patterned like a dead leaf and has a very short tail.

BELOW | Many chameleons exhibit unusual nasal protuberances, such as this male Strange-nosed Chameleon (*Kinyongia xenorhina*), which is endemic to the Rwenzori Mountains of Uganda and the Democratic Republic of Congo.

LEFT | The Namaqua Chameleon (*Chamaeleo namaquensis*), the "bulldog of chameleons," is a terrestrial species that preys on anything it can overpower.

coastal forests of Tanzania and Kenya by a different pygmy chameleon genus, *Rieppeleon*. The Great Lakes region of East Africa is also home to species in the genus *Kinyongia*, many of which exhibit strange nasal protuberances, while a seventh genus, *Nadzikambia*, containing only two species, is confined to Mozambique and Malawi.

The popular image of a chameleon is of a tree-dwelling lizard that feeds on insects, which it captures with its extendable sticky tongue, but one species above all others defies this stereotype. The Namaqua Chameleon (*C. namaquensis*) inhabits deserts and other arid habitats from Angola south to the Cape. It is predominantly a terrestrial species that marches across the ground in search of large insects and anything else it can overpower, which may include other vertebrates, from geckos to venomous snakes. The Namaqua Chameleon is truly the bulldog of chameleons.

CHAMAELEONIDAE
MADAGASCAN CHAMELEONS

ABOVE | The largest chameleon species is Parson's Chameleon (*Calumma parsonii*), a widely distributed species in northern and eastern Madagascar, found in rainforest habitats, especially along streams.

ABOVE | The Nosy Hara Leaf Chameleon (*Brookesia micra*) is the smallest known chameleon species, the smallest lizard, and the smallest vertebrate species.

Madagascar is home to both the largest chameleon in the world, Parson's Chameleon (*Calumma parsonii*, SVL 11⅔ in/ 295 mm), and the smallest, the Nosy Hara Dwarf Chameleon (*Brookesia micra*, SVL ¾ in/20 mm), from Nosy Hara, an island off the north coast. There are a further 94 species of Madagascan chameleons, many very localized and vulnerable to habitat loss or collection for the pet trade.

The two main genera are *Calumma*, which contains Parson's Chameleon, and *Furcifer*, which includes the second largest species, Oustalet's Chameleon (*F. oustaleti*). In general, *Calumma* species show a preference for rainforests or mountains, while *Furcifer* are more commonly encountered in dry woodland or thorn forest. One stunningly patterned species is the large (SVL 9¾ in/250 mm) Panther Chameleon (*F. pardalis*), a highly adaptable species that can survive in human-altered habitats.

The island also contains much smaller, less visible species. The genus *Brookesia* contains 30 species of tiny, short-tailed leaf chameleons, with two further small species now placed in the related genus *Palleon*. Some, like the Antsingy Leaf Chameleon (*B. perarmata*), are adorned with spines and crests and resemble tiny armored dragons, but other species are unadorned and drably colored, to blend into the dead leaves and straw-like grass on which they dwell. Spending their days hunting insects in the forest-floor leaf litter, they climb aloft to sleep at night. Their tails are less prehensile than those of the larger chameleons, and they lack the ability to undergo vivid color changes, relying instead on crypsis to avoid detection.

ABOVE | The bizarre Antsingy Armoured Leaf Chameleon (*Brookesia perarmata*) is the largest *Brookesia* species (SVL 60 mm, TTL 110 mm). It inhabits dry forest in a small area of karst limestone in western Madagascar.

ABOVE | The Panther Chameleon (*Furcifer pardalis*) is a large and adaptable species that can utilize habitats altered by humans, and in such situations it may out-compete smaller, rare, and less adaptable species.

IGUANIDAE
CENTRAL & SOUTH AMERICAN IGUANAS

The Green Iguana (*Iguana iguana*) has long been considered to comprise two subspecies, the nominate form inhabiting South America, and a northern South America–Central American form (*I. i. rhinolopha*) which is distinguished by a small, raised nasal horn. However, it seems that the diversity within the genus *Iguana* may have been underestimated, as researchers believe there are a number of species and subspecies hidden within this familiar but understudied "species."

Green Iguanas are large (SVL 22¾ in/580 mm) and feed primarily on vegetation. They are highly arboreal, but equally at home on the ground or in water, where they swim with ease, powered by their strong muscular tails. Adult males are impressive, with huge heads, large dewlaps for displaying, and orange shoulders during the breeding season. They engage in elaborate territorial displays to attract mates and deter rivals, and if they feel threatened will bite and use their long tails as whips.

DISTRIBUTION
North, South, and Central America, West Indies, Galapagos, and Fijian Islands

GENERA
Amblyrhynchus, Brachylophus, Cachryx, Conolophus, Ctenosaura, Cyclura, Dipsosaurus, Iguana, and *Sauromalus*

HABITATS
Rainforest, dry woodland, desert, and arid islands

SIZE
SVL 5 in (129mm) Santa Catalina Desert Iguana (*Dipsosaurus catalinensis*) to 29½ in (750 mm) Cayman Islands Iguana (*Cyclura nubila*)

ACTIVITY
Arboreal or terrestrial; diurnal

LEFT | A large male Green Iguana (*Iguana iguana*) of the Central American subspecies (*rhinolopha*), perched on a vantage point.

ABOVE | The Common Spiny-tailed Iguana (*Ctenosaura similis*) is a widely distributed Central American species, which also enters southern Mexico and is introduced to Florida.

RIGHT | The Campeche Spiny-tailed Iguana (*Cachryx alfredschmidti*) is found in southern Mexico and Guatemala and is listed as Near Threatened by the IUCN.

The other iguanas of Central America are the spinytail iguanas, belonging to the genera *Ctenosaura* (15 species from northwest Mexico to Panama) and *Cachryx* (two species from the Yucatán Peninsula).

REPRODUCTION
All genera are oviparous, the number of eggs laid depending on species and the size of the female

DIET
Primarily vegetarian, but juveniles often eat invertebrates and adults may scavenge carrion

According to recent research, the genus *Cachryx* is closely related to the Galapagos iguanas. As their name suggests, these iguanas possess extremely spiny tails. Generally black to gray, they are also smaller than Green Iguanas, the largest being the Black Spinytail Iguana (*Ctenosaura similis*), which may achieve 19¼ in (490 mm) SVL. Several species are listed as Critically Endangered to Vulnerable by the IUCN, including the Utila Island Spinytail Iguana (*Ctenosaura bakeri*) and Black-chested Spinytail Iguana (*C. melanosterna*).

IGUANIDAE
WEST INDIAN IGUANAS

BELOW | The Grand Cayman Blue Iguana (*Cyclura lewisi*) came back from the brink of extinction, thanks to an intensive captive breeding and education program.

The West Indian iguanas (*Cyclura*) inhabit the Greater Antilles, the Bahamas, and the Virgin Islands. They are diurnally active inhabitants of xerophytic cactus scrub and mesophytic habitats, usually in lowland rocky locations, especially near the coast. They are primarily vegetarian, but adults will scavenge carrion.

These impressive lizards are among the most rapidly declining of all reptile groups. Two species, from St. Thomas and Puerto Rico, disappeared during pre-Columbian times, and the Navassa Rhinoceros Iguana (*C. onchiopsis*) followed them in the late nineteenth century, hunted by miners. The Jamaican Iguana (*C. collei*) survives as a population of 150 in the Hellshire Hills, Jamaica, and the Grand Cayman Blue Iguana (*C. lewisi*) was on the brink of extinction, with 50 specimens left in the wild, until it became the subject of an intensive conservation program. Its population continued to fall to between five and 15 individuals by 2003, but captive breeding and release programs have now increased the wild population considerably. Ongoing threats include rats foraging eggs, mongooses and cats killing juveniles, and goats or cactus moths devouring the vegetation the adult iguanas rely upon.

Ten extant *Cyclura* species are recognized, four with subspecies. The largest is the Cuban Iguana (*C. nubila nubila*), males achieving $29\frac{1}{3}$ in (745 mm) SVL, but the best known is the heavily built and aptly named Hispaniolan Rhinoceros Iguana (*C. cornuta*), easily recognized by the raised pseudo-horns on its snout. The smallest are the Turks and Caicos Iguana (*C. carinata*) and the Central Bahamian Rock Iguana (*C. rileyi*), adult females of which are less than $11\frac{3}{4}$ in (300 mm) SVL, making them vulnerable to invasive predators.

The Lesser Antillean Iguana (*Iguana delicatissima*) is a close relative of the well-known Green Iguana (*I. iguana*) of the Latin American mainland, from which it can easily be distinguished because it lacks the latter's large, distinctive subtympanic plates. Diurnal and arboreal, it is also primarily vegetarian.

LEFT | The Rhinoceros Iguana (*Cyclura cornuta*) from Hispaniola, is well named because of the large pseudo-horns on its snout.

BELOW | The Lesser Antillean Green Iguana (*Iguana delicatissima*) can be distinguished from its more famous relative, the Green Iguana (page 184) because it lacks the large white subtympanic plates on the back of the neck.

IGUANIDAE
NORTH AMERICAN IGUANAS & CHUCKWALLAS

The smallest true iguanas are the desert iguanas (*Dipsosaurus*), which inhabit southwestern USA and northwestern Mexico. There are two species recognized, most of the range being occupied by the Desert Iguana (*D. dorsalis*), while the population from Santa Catalina Island in the Sea of Cortez is treated as a distinct species, the Santa Catalina Desert Iguana (*D. catalinensis*). Neither species exceeds 6 in (154 mm) SVL. The habitat of the Desert Iguana is largely xerophytic semidesert with creosote bushes, under which they may construct their burrows, running to them if danger threatens, but they are also found in more

ABOVE | The smallest species of iguana are the desert iguanas. The Desert Iguana (*Dipsosaurus dorsalis*) inhabits soutwestern USA and northwestern Mexico.

RIGHT | The Northern or Common Chuckwalla (*Sauromalus ater*) is the most widely distributed chuckwalla, occurring as five subspecies on the mainland. The other four species are found on islands in the Gulf of California.

IGUANIA—Agamas, chameleons, and iguanas

mesophytic habitats in the south of their range. Desert iguanas are primarily vegetarian, climbing into bushes to feed on fruit, buds, and flowers, but they will also eat insects, carrion, or their own fecal pellets, to maintain the microbial gut flora necessary to digest leaves and other vegetation. These iguanas are reported to be more heat tolerant than many other lizards, basking or remaining active in the heat of the day.

The chuckwallas are large, rotund lizards. The widest distributed species is the Northern or Common Chuckwalla, *Sauromalus ater*. The range of the five species largely overlaps that of the Desert Iguana: southwestern USA (California, Arizona, Nevada, Utah) and northwestern Mexico (Sonora, Sinaloa, and Baja California). As with the Desert Iguana, some of the island populations in the Sea of Cortez are given specific status, such as the Santa Catalina Chuckwalla (*S. klauberi*) and San Esteban Chuckwalla (*S. varius*). Chuckwallas are almost entirely herbivorous, although a few insects are also eaten. They inhabit rocky outcrops in creosote desert scrub and retreat to rocky crevices at night or if threatened. They will attempt to prevent extrication by gulping air to make themselves fit their retreats more tightly.

ABOVE | The Piebald or San Esteban Chuckwalla (*Sauromalus varius*) is endemic to San Esteban Island, and neighboring islets, in the Gulf of California.

IGUANIDAE
GALAPAGOS & FIJIAN IGUANAS

Not all iguanas are found on the American mainland or on Caribbean islands. The Galapagos Archipelago, 600 miles (965 km) west of the South American mainland, is home to four species of remarkable iguanas. The most famous is the Marine Iguana (*Amblyrhynchus cristatus*), which is divided into 12 island subspecies. This large lizard (< 22 in/560 mm SVL) is heavily built, black, dark gray, or sometimes reddish-pink, and feeds entirely on seaweed, the males diving deep into the cold waters of the Humboldt Current to obtain a meal, before returning to the black lava rocks to bask. Females and juveniles feed in the intertidal zone at low tide.

BELOW | Male Marine Iguanas (*Amblyrhychus cristatus*) bask to warm up when they return to land after diving deep into the cold Humboldt Current to feed on red and brown seaweed.

IGUANIA—Agamas, chameleons, and iguanas

Less famous are the three species of land iguanas (*Conolophus*). Until recently, only two species were recognized, the widespread Galapagos Land Iguana (*C. subcristatus*), which occurs on several islands, and the Santa Fe Land Iguana (*C. pallidus*). However, in 2009 a third species, the Isabela Island Pink Land Iguana (*C. marthae*), was described. Land iguanas eat prickly pear cactus, which also provides them with their water supply on the arid, volcanic Galapagos Islands.

The most remote true iguanas are those that live over 6,210 miles (10,000 km) farther west, on the Fijian Islands. Four extant species are recognized: the widespread Fijian Banded Iguana (*Brachylophus bulabula*); the Lau Banded Iguana (*B. fasciatus*), from the Lau Islands, and also introduced to Tonga; the Fijian Crested Iguana (*B. vitiensis*), also relatively widespread; and the recently described Gau (pronounced "Ngau") Banded Iguana (*B. gau*), from Gau Island. Fijian iguanas are relatively small compared to their American relatives, bright green, herbivorous, and highly arboreal. Their ancestors are thought to have been transported across the Pacific on rafts of red mangrove carried by ocean currents.

All Galapagos and Fijian iguanas are listed as Endangered or Critically Endangered by the IUCN.

ABOVE | The Santa Fe Land Iguana (*Conolophus pallidus*) is one of three species of land iguanas in the Galapagos Islands. Unlike the Marine Iguana they eat cactus fruit and pads.

BELOW | The brilliantly verdant Fijian Banded Iguana (*Brachylophus fasciatus*) is one of four iguanas inhabiting the Fijian Islands over 6,200 miles (10,000 km) west of the Americas.

LEIOCEPHALIDAE
CURLYTAILS

The Leiocephalidae contains a single genus, *Leiocephalus*, which has speciated widely in the Caribbean. Twenty-four extant species are recognized, plus 68 subspecies, from all the major islands and archipelagos. Six further species are only known from fossils (Hispaniola, Jamaica, Puerto Rico, Antigua, and Barbuda), and three species have gone extinct in historical times (Hispaniola, Puerto Rico, and Martinique), presumably through the actions of humans. The IUCN lists five species as Critically Endangered, one as Endangered, and two as Vulnerable, a reflection of their island endemicity. Two of the more adaptable species, the Northern Curlytail (*L. carinatus*) and Red-sided Curlytail (*L. schreibersii*), have been introduced to Florida.

The common name "curlytail" is descriptive. When the lizards feel threatened they curl their tails over their backs, possibly to mimic a scorpion. These fast-moving, diurnal lizards feed on invertebrates but will also take smaller lizards, and occasionally fruit and flowers. Males are larger than females.

LEFT | The Northern Curly-tail Lizard (*Leiocephalus carinatus*) is the commonest species with 12 subspecies across Cuba, the Cayman Islands, and the Bahamas.

RIGHT | A short, prickly tail gives the Weapontail (*Hoplocercus spinosus*) its common name. It inhabits arid Cerrado habitats in central Brazil and Bolivia.

DISTRIBUTION
West Indies: Cuba, Hispaniola, Puerto Rico, Cayman Islands, Navassa, Bahamas, Turks and Caicos, Lesser Antilles, and Honduras (Swan Islands); introduced to Florida

GENUS
Leiocephalus

HABITATS
Dry woodland, rocky areas, cacti scrub, and beaches

SIZE
SVL 2 in (53 mm) Hispaniolan Pale-bellied Curlytail (*Leiocephalus semilineatus*) to 5¼ in (133 mm) Saw-scaled Curlytail (*L. carinatus*); extinct species were larger, up to 7¾ in/200 mm (Leeward Islands Curlytail, *L. cuneus*)

ACTIVITY
Terrestrial; diurnal and heliophilic

REPRODUCTION
Oviparous, females laying two eggs

DIET
Insects, spiders, smaller lizards, and fruit

HOPLOCERCIDAE
WOODLIZARDS, MANTICORES & WEAPONTAIL

The Hoplocercidae is primarily a northern South American genus that also occurs on the Pacific versant from Ecuador to Panama. The woodlizards (*Enyalioides*) and manticores (*Morunasaurus*) are inhabitants of rainforest, but the Weapontail (*Hoplocercus spinosus*) inhabits the dry Cerrado of central Brazil and Bolivia. The semi-arboreal woodlizards are somewhat laterally compressed, with a crest running off the head and down the back, resembling Asian dragons (*Japalura*, page 86), while the terrestrial manticores and the Weapontail are more dorsoventrally compressed with spiny tails, and resemble dhab lizards (*Uromastyx*, page 176). The Weapontail has a very short, excessively spiny tail and also spines on the back.

Manticores and the Weapontail are believed to excavate their own burrows or shelter in rocky crevices, while the woodlizards may either sleep aloft or retreat to burrows on the ground.

DISTRIBUTION
Southern Central America and northern South America, from Panama to Brazil and Bolivia

GENERA
Enyalioides, *Hoplocercus*, and *Morunasaurus*

HABITATS
Rainforest and Cerrado

SIZE
SVL 3¾ in (96 mm) Cordillera Azul Woodlizard (*Enyalioides azulae*) to 7½ in (192 mm) Red-eyed Woodlizard (*E. oshaughnessyi*)

ACTIVITY
Semi-arboreal or terrestrial; diurnal, but some species are heliophobic (*Hoplocercus*)

REPRODUCTION
All species are oviparous, but little is known; six eggs are reported for the Broad-headed Woodlizard (*Enyalioides laticeps*)

DIET
Insects and spiders

CROTAPHYTIDAE
COLLARED & LEOPARD LIZARDS

Crotaphytidae is a North American family centered on the American Southwest, but also occurring in central USA, and south into northern Mexico, especially Baja California. The collared lizards (*Crotaphytus*) are diurnal inhabitants of arid open habitats, where they occupy elevated vantage points to watch for prey, and presumably also predators. They feed primarily on arthropods, although large individuals will also capture other lizards, and even rodents and small snakes, running the prey down and using their powerful jaws to crush it to death. This powerful bite is also utilized in territorial disputes between rival males.

The leopard lizards (*Gambelia*) are also diurnal, but are more sit-and-wait ambushers that hunt in scrubby vegetation, again feeding primarily on arthropods, with occasional smaller lizards included.

Leopard lizards are capable of caudal autotomy to escape a predator, but collared lizards do not possess the necessary fracture planes in their tail vertebrae. They are, however, able to escape using a curious bipedal jumping motion to flee across rocks.

DISTRIBUTION
Central and southwestern USA, and northern Mexico

GENERA
Crotaphytus and *Gambelia*

HABITATS
Open desert, semidesert, or rocky outcrops

SIZE
SVL 4 in (99 mm) Grismer's Collared Lizard (*Crotaphytus grismeri*) to 5¾ in (146 mm) Long-nosed Leopard Lizard (*Gambelia wislizenii*)

ACTIVITY
Terrestrial or saxicolous; diurnal and heliophilic

REPRODUCTION
All species are oviparous, with clutch sizes of 3–8 eggs

DIET
Primarily arthropods, but large individuals also take small vertebrates, including small reptiles and rodents

CORYTOPHANIDAE
BASILISKS & CASQUE-HEADED LIZARDS

The Corytophanidae contains three genera and 11 species, found in forested habitats and along rivers from Mexico to Ecuador and Venezuela, with the greatest number of species (nine) occurring in Honduras. The basilisks, of genus *Basiliscus*, are also known as "Jesus Christ lizards" because of their ability to run across water (see page 41). Juveniles or adults of the smallest species, the Brown Basilisk (*B. vittatus*), make the most proficient use of this escape strategy. The helmeted basilisks (*Corytophanes*) and casque-headed basilisks (*Laemanctus*) rely more on crypsis to escape detection.

The members of this family are noted for their impressive head adornments, but these are sexually dimorphic in *Basiliscus*, only males achieving large crests, while both sexes are in possession of crests or casques in the other two genera.

It is often reported that all members of the Corytophanidae are oviparous, but the smallest species, the Keeled Helmeted Basilisk (*C. percarinatus*), is viviparous.

LEFT | The Collared Lizard (*Crotaphytus collaris*) is a powerful lizard that often takes down smaller lizards as prey.

RIGHT | The Smooth Helmeted Basilisk (*Corytophanes cristatus*) is a secretive forest species that prefers to stay motionless and hide rather than run from a threat.

DISTRIBUTION
North and Central America to northwest South America

GENERA
Basiliscus, *Corytophanes*, and *Laemanctus*

HABITATS
Rainforest, especially near water; dry scrub

SIZE
SVL 4⅓ in (110 mm) Keeled Helmeted Basilisk (*Corytophanes percarinatus*) to 9¾ in (250 mm) Plumed Basilisk (*Basiliscus plumifrons*) and Common Basilisk (*B. basiliscus*)

ACTIVITY
Highly arboreal, although *Basiliscus* is equally at home on the ground; diurnal

REPRODUCTION
All species are oviparous, except *C. percarinatus*, which is viviparous and produces up to seven neonates

DIET
Insects and spiders

LEIOSAURIDAE—ENYALIINAE & LEIOSAURINAE
SOUTH AMERICAN TREE & GROUND LIZARDS

The family Leiosauridae contains two subfamilies. The Enyaliinae are slender lizards with long, slender tails and limbs, suggesting an agile arboreal lifestyle, while the Leiosaurinae contains big-headed, stouter lizards, with shorter limbs, which are terrestrial in habit.

Within the Enyaliinae the Fathead Anoles (*Enyalius*) are slow-moving arboreal inhabitants of the Amazonian and Atlantic coastal forests of Brazil, with one species entering Uruguay. They change colour from brown at rest to green when active. The equally arboreal Tree Lizards (*Anisolepis*) inhabit both rainforest and seasonally-flooded grasslands, while the Steppe Lizards (*Urostrophus*) are more terrestrial, inhabiting arid semidesert.

The Leiosaurinae contains the stout-bodied Patagonian Grumblers (*Diplolaemus*), which occur from Patagonia to the Strait of Magellan, in the far south of Argentina and Chile. The Southern Anoles (*Leiosaurus*) are endemic to Argentina, while the Big-headed Anoles (*Pristidactylus*), contains

LEFT | Ihering's Fatheaded Para-anole (*Enyalius iheringii*) is an arboreal inhabitant of the southern Atlantic Coastal Forests of Brazil. It may be green or gray in color.

ENYALIINAE

DISTRIBUTION (BLUE ON MAP)
South America, from Amazonian and Atlantic coastal Brazil, to Bolivia, Paraguay, Uruguay, and Argentina

GENERA
Anisolepis, *Enyalius*, and *Urostrophus*

HABITATS
Tropical rainforest, southern Chaco forest, seasonally flooded grassland, and arid semidesert

SIZE
SVL 3 in (78 mm) Argentine Steppe Lizard (*Urostrophus gallardoi*) to 5 in (124 mm) Ihering's Fathead Para-anole (*Enyalius iheringii*)

ACTIVITY
Primarily arboreal, or occasionally terrestrial; diurnal

REPRODUCTION
All species are oviparous; clutch size is largely unknown, possibly up to 15 eggs

both Argentine and Chilean endemics. It is curious, given their southern distribution, that these species are oviparous when squamate reptiles living at such latitudes might be expected to be viviparous.

Sometimes the term "para-anoles" is used for these lizards, to distinguish them from the true anoles of the Dactyloidae (page 200).

RIGHT | Darwin's Grumbler (*Diplolaemus darwinii*) is a stocky terrestrial lizard from Patagonian Argentina and Chile.

DIET
Arthropods, primarily insects

LEIOSAURINAE

DISTRIBUTION (RED ON MAP)
South America, from southern Amazonian Brazil to Argentina and Chile

GENERA
Diplolaemus, *Leiosaurus*, and *Pristidactylus*

HABITATS
Southern beech forests

SIZE
SVL 2¾ in (73 mm) Casuhatien Big-headed Para-anole (*Pristidactylus casuhatiensis*) to 4¾ in (120 mm) Catamarca Para-anole (*Leiosaurus catamarcensis*)

ACTIVITY
Primarily terrestrial, or possibly arboreal; diurnal

REPRODUCTION
All species are oviparous; clutch size is largely unknown, but possibly up to 15 eggs

DIET
Arthropods, primarily insects

LIOLAEMIDAE
SOUTH AMERICAN SWIFTS & TREE LIZARDS

The Liolaemidae is a large family from Andean and Patagonian South America. It contains three genera, including the monotypic, oviparous Peruvian Desert Lizard (*Ctenoblepharys adspersa*) from the coastal dunes of Peru. The genus *Phymaturus* contains 47 species, commonly referred to as mountain lizards, which are distributed through the Andes of Argentina and Chile to elevations of 11,500 ft (3,500 m).

The third genus is *Liolaemus*, the third largest lizard genus in the world, with 273 species. Distributed across the southern half of South America, *Liolaemus* species are referred to as snow swifts or mountain, ground, or tree lizards, depending on their habitat preferences. There are both oviparous and viviparous species, the latter strategy being an adaptation for life in cooler climates. The Zodiac Lizard (*L. signifer*) occurs up to 14,100 ft (4,300 m) in the Peruvian Andes, while the Huaca-Huasi Lizard (*L. huacahuasicus*) achieves 13,100 ft (4,000 m) in Argentina. The two most southerly distributed lizards in the world belong here, Sarmiento de Gamboa's Lizard (*L. sarmientoi*) and the Strait of Magellan Lizard (*L. magellanicus*); these occur in sympatry on Tierra del Fuego.

DISTRIBUTION
Southern South America, from Peru to southern Argentina and Chile

GENERA
Ctenoblepharys, *Liolaemus*, and *Phymaturus*

HABITATS
Coastal desert, open grasslands, dry woodlands, and rocky mountain slopes

SIZE
SVL 2 in (50 mm) Diaguita Lizard (*Liolaemus diaguita*) to 4¾ in (120 mm) Hooded Mountain Lizard (*Phymaturus verdugo*)

ACTIVITY
Terrestrial, arboreal, or saxicolous; diurnal

REPRODUCTION
Ctenoblepharys is oviparous, *Phymaturus* is viviparous, and *Liolaemus* contains both oviparous and viviparous species. The El Nihuil Lizard (*Liolaemus parthenos*) is a parthenogenetic oviparous species

DIET
Small invertebrates, including insects and spiders, and occasional fruit and flowers

POLYCHROTIDAE
BUSH PARA-ANOLES

The Polychrotidae contains a single genus, *Polychrus*, with eight species. They are generally larger and of stockier build than the true anoles of genus *Anolis* (page 200), and are called "para-anoles" here to distinguish them from the true anoles. The best-known species is the Marbled Bush Para-anole (*P. marmoratus*), from northeastern South America. This generally bright green species moves through the vegetation in a slow, ponderous way, grasping the vegetation with its feet and tail in a manner reminiscent of an Old World chameleon, and earning the colloquial name of "monkey lizard." It can also undergo color changes, from green to brown, dependent on mood, and remain stationary for long periods, waiting to ambush large insects. The northernmost species is the Central American Bush Para-anole (*P. gutturosus*), occurring from Honduras to Ecuador, while the southernmost is the Brazilian Bush Para-anole (*P. acutirostris*), which extends into northern Argentina.

LEFT | Fabian's Lizard (*Liolaemus fabiani*) is a swift from the Atacama Desert of Chile.

RIGHT | The Marbled Bush Para-anole (*Polychrus marmoratus*) is the best known member of the genus *Polychrus*.

DISTRIBUTION
Central and South America, from Honduras to Brazil

GENUS
Polychrus

HABITATS
Rainforest, especially secondary growth with numerous low bushes and shrubs

SIZE
SVL 4¼ in (108 mm) Ecuadorian Bush Para-anole (*Polychrus femoralis*) to 6⅔ in (170 mm) Central American Bush Para-anole (*P. gutturosus*)

ACTIVITY
Highly arboreal sit-and-wait ambushers; diurnal

REPRODUCTION
All species are oviparous, producing 7–31 eggs (Sharp-nosed Bush Para-anole, *P. acutirostris*)

DIET
Arthropods, primarily large insects, but also leaves, flowers, and fruit

DACTYLOIDAE
TRUE ANOLES

The genus *Anolis* is the largest reptile genus in the world. As currently recognized, it contains 437 species and an additional 102 subspecies. There are moves to split this huge genus into up to eight separate genera, but this is controversial.

Despite the enormous diversity in the anoles, there is only one native continental United States species, the North American Green Anole (*Anolis carolinensis*), which is found from North Carolina to Texas. Of Florida's 12 species, 11 are introduced from the West Indies, while the North American Green Anole has itself been introduced to Hawaii, Japan, and Saipan in Micronesia. It is believed to be the most studied lizard in the world, and it was the first reptile to have its entire genome sequenced.

At the other end of the range, the southernmost anole is the Southern Anole (*A. meridionalis*), which occurs as far south as Paraguay and southern Brazil.

Anoles are agile climbers, most being arboreal green or brown inhabitants of trees and bushes, but

DISTRIBUTION
Southeastern North America, Central America, northern and central South America, and the West Indies

GENUS
Anolis

HABITATS
Rainforest, woodland, gardens, grasslands, and islands

SIZE
SVL 1⅓ in (34 mm) Stripe-bellied Grass Anole (*Anolis cupeyalensis*) to 7½ in (191 mm) Western Giant Anole (*A. luteogularis*)

ACTIVITY
Arboreal; diurnal

REPRODUCTION
All species are oviparous, females multi-clutching, laying clutches of 1–2 eggs on several occasions each year

DIET
Arthropods, especially insects; larger species also take smaller lizards, including congeners; some species eat flowers and fruit

there are also numerous grassland species, such as the Common Grass Anole (*A. auratus*), which is striped with yellow to help it blend into its linear environment. An anole's excellent climbing ability is due to the presence of specialized lamellae under the toes that are similar to those possessed by geckos.

Often several different species of anoles will occur in sympatry, so males of each species exhibit differently colored dewlaps under the throat, with which they signal their presence to attract a mate. These dewlaps provide important clues to biologists in the identification of species.

The diversity of anoles, especially in the Caribbean and Central America, has led to them being used for numerous biogeographical, behavioral, ecological, and evolutionary studies.

BELOW | The Common Grass Anole (*Anolis auratus*), of southern Central and northern South America, is striped to blend in with its linear environment.

LEFT | The Green Anole (*Anolis carolinensis*) is the only native species from the 12 found in Florida. It is also the most studied lizard in the world.

BELOW | The Horned Anole (*Anolis proboscis*), from Ecuador, is sexually dimorphic, the male (left) bearing a large nasal projection to rival that of any chameleon.

PHRYNOSOMATIDAE—PHRYNOSOMATINAE & SCELOPORINAE

HORNED LIZARDS, SPINY LIZARDS & THEIR ALLIES

The genus *Phrynosoma* contains 21 species of horned lizards. These rotund but flattened-bodied lizards inhabit deserts and arid grasslands; their distribution is centered on southwestern USA and Mexico, with two species just entering Canada, and the Giant Horned Lizard (*P. asio*) occurring as far south as Guatemala. Horned lizards are cryptically patterned to match the substrate and if they feel threatened they will freeze, hugging the ground and blending in perfectly. The body outline is broken by numerous spiny scales and the back often bears rows of spinous scales, but it is the head that is most impressive, resembling that of a miniature dragon,

BELOW LEFT | The Texas Horned Lizard (*Phrynosoma cornutum*) blends in beautifully with its desert habitat as it spends the day eating ants.

RIGHT | The male Emerald Swift (*Sceloporus malachiticus*) is bright green; females are more drably patterned. It lives at high elevations in Central America.

with backward-pointing spikes. Horned lizards feed primarily on ants, and possess a remarkable defensive tactic of squirting narrow streams of blood from their eyes at a perceived predator.

Also in the Phrynosomatinae, in the American deserts, Zebra-tailed Lizards (*Callisaurus draconoides*) dash to cover, their tails curled forward over their bodies, exposing the black and white banded undersides. The Greater Earless Lizard (*Cophosaurus texanus*) and lesser earless lizards (*Holbrookia*) inhabit dry grasslands and rocky slopes, while fringe-toed lizards (*Uma*) have specialized toes for running across loose desert sand. A seventh species of genus *Uma* was described in 2020, with the name *Uma thurmanae*, the -ae suffix indicating it was named for a female. Even taxonomists like the occasional joke!

In the Sceloporinae, the genus *Sceloporus* contains 108 species, from Canada to Panama, bearing names based on where they occur, such as fence, prairie, canyon, plateau, or sagebrush lizards. Some

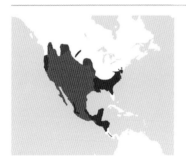

PHRYNOSOMATINAE

DISTRIBUTION (RED AND PURPLE ON MAP)
North and Central America, from Canada to Guatemala

GENERA
Callisaurus, *Cophosaurus*, *Holbrookia*, *Phrynosoma*, and *Uma*

HABITATS
Desert, arid grassland, and rocky outcrops

SIZE
SVL 2½ in (62 mm) Keeled Earless Lizard (*Holbrookia propinqua*) to 5 in (130 mm) Texas Horned Lizard (*Phrynosoma cornutum*)

ACTIVITY
Terrestrial or saxicolous; diurnal and heliophilic

REPRODUCTION
Most species are oviparous, but *Phrynosoma* contains both oviparous and viviparous species

are extremely spiny, others smooth-scaled. The Emerald Swift (*S. malachiticus*) is a high-elevation Central American species (< 9,190 ft/2,800 m) inhabiting rocky outcrops in cloud forest. Males are vivid green with blue and black markings. *Sceloporus* from lower elevations are oviparous, but this species is viviparous in response to its high-elevation home.

The Side-blotched Lizard (*Uta stansburiana*) occurs across western USA and Mexico, and feeds on a range of different arthropods. The side-blotched lizard population on Isla Coloradito in the Gulf of California, the Swollen-snouted Side-blotched Lizard (*U. tumidarostra*), has a hugely swollen snout. There are no insects on the small island, only a sea lion colony, and this species has evolved to feed solely on the marine isopods in the intertidal zone. These small lizards can tolerate higher levels of salinity in their diet than even the Galapagos Marine Iguana (*Amblyrhynchus cristatus*, page 184), expelling excess salt by physically snorting it out from their enlarged nostrils.

The saxicolous rock lizards (*Petrosaurus*) and arboreal tree lizards (*Urosaurus*) live in rocky habitats, the latter of those with plentiful vegetation.

DIET
Arthropods, primarily ants in *Phrynosoma*, and also beetles; some species, such as *Callisaurus*, also eat flowers and buds

SCELOPORINAE

DISTRIBUTION (BLUE AND PURPLE ON MAP)
North and Central America, from Canada to Panama

GENERA
Petrosaurus, Sceloporus, Urosaurus, and *Uta*

HABITATS
Dry forest, upland pine forest, boulder piles, canyons, fence lines, drystone walls, grasslands, stony deserts, and rocky islands

SIZE
SVL 2 in (50 mm) Spotted Spiny Lizard (*Sceloporus maculosus*) to 7 in (175 mm) Baja California Rock Lizard (*Petrosaurus thalassinus*)

ACTIVITY
Terrestrial, saxicolous, or arboreal (*Urosaurus*); diurnal and heliophilic

REPRODUCTION
Most species are oviparous, but *Sceloporus* also contains viviparous species from higher elevations

DIET
Arthropods, primarily insects, although some species also eat flowers and buds

TROPIDURIDAE
LAVA LIZARDS, WHORLTAILS & TREERUNNERS

The Tropiduridae comprises seven South American and one Galapagos genera. The Galapagos lava lizards (*Microlophus*) and the mainland lava lizards (*Tropidurus*) are sun-lovers that bask on black laval rocks, as do the endemic Brazilian lava lizards (*Eurolophosaurus*) and Atlantic Forest Lava Lizard (*Strobilurus torquatus*). The largest genus is *Stenocercus*, the whorltail lizards, terrestrial lizards with spiny dorsal crests and tails.

Tropiduridae also contains tree-dwelling species such as the thornytail lizards (*Uracentron*), which feed on tree ants and often live in polygynous groups, with one male and several females sharing a single tree. The South American, slender, long-legged, and blunt-headed Mop-headed Lizard (*Uranoscodon superciliosus*) is an arboreal riverbank and flooded forest inhabitant that runs across the water in the same way as the basilisks (*Basiliscus*, page 195).

RIGHT | The Green Thornytail Lizard (*Urocentron azureum*) is a spiny-tailed tree dweller. The word *azureum* means "blue" so the author who described this lizard never saw a live one: green pigment goes blue in preservative.

DISTRIBUTION
South America and the Galapagos Islands

GENERA
Eurolophosaurus, Microlophus, Plica, Stenocercus, Strobilurus, Tropidurus, Uracentron, and *Uranoscodon*

HABITATS
Rainforest, rocky habitats, and oceanic islands

SIZE
SVL 2¼ in (57 mm) Huarmey Whorltail Lizard (*Stenocercus johaberfellneri*) to 7 in (177 mm) Common Treerunner (*Plica plica*)

ACTIVITY
Arboreal, terrestrial, or saxicolous; diurnal and heliophilic

REPRODUCTION
All species are oviparous

DIET
Arthropods, especially large tree ants (*Plica, Uracentron, Strobilurus*); also some fruit and flowers (*Tropidurus*)

OPLURIDAE
MALAGASY SWIFTS & IGUANIANS

BELOW | The Western Malagasy Swift (*Oplurus cuvieri*), a distinctive lizard with a black and white collar, is the only member of the Opluridae to also occur outside Madagascar, in the Comoro Islands.

The Opluridae is the only pleurodont family to occur entirely outside the Americas. It comprises two small genera and eight species, seven of which are endemic to Madagascar, while Cuvier's Malagasy Swift (*Oplurus cuvieri*) also occurs on Grande Comore Island to the north. This species is almost indistinguishable from Merrem's Malagasy Swift (*O. cyclurus*); they both exhibit black and white collars and dorsal markings, and both have extremely spiny tails, but the tail of Cuvier's Malagasy Swift has a row of small scales between the spiny whorls, which is absent in Merrem's Malagasy Swift. The other four species of *Oplurus* have less spiny tails.

The second genus is *Chalarodon*, usually referred to as Madagascan iguanians, although both species are smaller than the swifts in *Oplurus*. Both genera have a preference for the arid habitats of southern and western Madagascar, and are not found in the wet forests of the east and north. The Malagasy swifts are arboreal or saxicolous, inhabiting trees or rocky outcrops, while the Madagascan iguanians are terrestrial. Male Madagascan iguanians have a raised crest from the back of the head to the tail.

DISTRIBUTION
Western and southern Madagascar, and Grande Comoro Island

GENERA
Chalarodon and *Oplurus*

HABITATS
Dry forests, rocky outcrops, degraded monsoon forest, or plantation clearings

SIZE
SVL 2⅓ in (60 mm) Steinkamp's Madagascar Iguanian (*Chalarodon steinkampi*) to 6⅓ in (160 mm) Merrem's Madagascar Swift (*Oplurus cyclurus*)

ACTIVITY
Terrestrial, arboreal, or saxicolous; diurnal

REPRODUCTION
All species are oviparous, laying clutches of two (*Chalarodon*) or 2–6 (*Oplurus*) eggs

DIET
Arthropods

LEFT | The slow-moving Chinese Crocodile Lizard (*Shinisaurus crocodilurus*) is the sole member of its family (Shinisauridae) and is threatened by habitat loss within its small range.

INFRAORDER ANGUIMORPHA

The infraorder Anguimorpha contains seven families of lizards. Many exhibit body elongation and limb loss, combined with secretive, semi-fossorial lifestyles, but others are among the most visible, alert, agile, and voracious lizards alive today. The American anguimorph families and subfamilies are Anniellinae (Californian legless lizards), Gerrhonotinae (alligator lizards), Diploglossidae (galliwasps and South American legless lizards), Xenosauridae (knob-scaled lizards), and Helodermatidae (Gila Monster and beaded lizards).

The subfamily Anguinae (slow worms and glass lizards) is also present in North America, as well as in Europe, North Africa, and Southeast Asia, while the fully Asian families are the Shinisauridae (Chinese Crocodile Lizard) and Lanthanotidae (Borneo Earless Monitor). The largest family is the Varanidae (monitor lizards), which is distributed across Africa, Arabia, Asia, and Australasia. This family contains the largest living, and the largest ever known, lizard species.

The helodermatids were for a long time considered to be the only venomous lizards, but many members of the Anguimorpha are now believed to possess oral venom glands, and bites from large varanid lizards can be very serious.

It is from within the Anguimorpha that snakes are thought to have evolved.

ANGUIDAE—ANGUINAE
SLOW WORMS & GLASS LIZARDS

Subfamily Anguinae contains five genera of elongate, limbless lizards that may be mistaken for snakes. The largest species, the Scheltopusik (*Pseudopus apodus*), from southeastern Europe and Western Asia, has been known as the "glass snake" due to its snake-like appearance and because it occasionally practices caudal autotomy (see page 73), breaking off its own tail to escape predation.

This trait is more frequently observed in the Slow Worm (*Anguis fragilis*, which translates as "fragile snake"). Despite appearing to be legless, the Scheltopusik does possess a pair of vestigial, flap-like hind limbs, and also external ear openings, and both it and the five species of European slow worms (*Anguis*) exhibit the distinctly non-snake-like characteristic of movable eyelids.

Glass lizards are not confined to Europe and Western Asia. Seven species of Asian glass lizards (*Dopasia*) are found in tropical forests, from Assam to southern China, and on the islands of Hainan, Taiwan, Sumatra, and Borneo, while the Moroccan

LEFT | The Slow Worm (*Anguis fragilis*) is neither slow, nor a worm, nor a snake. Unlike the Scheltopusik opposite, it lacks external ear-openings and hindlimbs as vestigial flaps, but it still demonstrates its lizard credentials with the possession of eyelids and a notched, rather than forked, tongue.

DISTRIBUTION
Europe, western, Southern, and Southeast Asia, northwest Africa, eastern North America, and Mexico

GENERA
Anguis, Dopasia, Hyalosaurus, Ophisaurus, and *Pseudopus*

HABITATS
Temperate woodland, grasslands, Mediterranean maquis, and lowland or submontane tropical rainforest

SIZE
SVL 5 in (125 mm) Borneo Glass Lizard (*Dopasia buettikoferi*) to 22¾ in (580 mm) Scheltopusik (*Pseudopus apodus*)

ACTIVITY
Semi-fossorial or fossorial; diurnal, but secretive

ABOVE | With a total length of up to 50 in (1.3 m), mostly tail, the Scheltopusik (*Pseudopus apodus*) is the largest legless lizard in the world.

REPRODUCTION
All species are oviparous, laying 2–40 parchment-shelled eggs, except *Anguis*, which are viviparous, giving birth to 2–15 neonates

DIET
Earthworms, slugs, or arthropods, but larger species (*Pseudopus*) take snails, birds' eggs, fledgling birds, small snakes, or small mammals

Glass Lizard (*Hyalosaurus koellikeri*) occurs in wet wooded dunes in the Atlas Mountains of northwest Africa. Across the Atlantic, four species of American glass lizards (*Ophisaurus*) inhabit eastern and central USA, with two further species on the Caribbean coast of Mexico. The Eastern Glass Lizard (*O. ventralis*) has been introduced to the Cayman Islands.

Slow worms and the Asian and Moroccan glass lizards feed on invertebrates, but the Scheltopusik and the American glass lizards also eat small vertebrates, including mice, snakes, or birds' eggs. Their rigid bodies are heavily armored with plate-like scales, but a lateral groove along the flanks enables them to expand when swallowing more bulky prey, or when females are carrying eggs.

ANGUIDAE—GERRHONOTINAE
ALLIGATOR LIZARDS

LEFT | At some time in the past this Trans-volcanic Alligator Lizard (*Barisia imbricata*) has autotomized its tail to escape a threat, and subsequently regenerated a new but shorter tail.

The Gerrhonotinae are the anguids with properly formed, albeit shortened, pentadactyl limbs, and elongate bodies. It is thought that they may resemble the ancestral precursors of snakes, which are believed to have evolved from within the Anguimorpha. Alligator lizards are so-named because they possess armored heads and powerful jaws, and are covered in plate-like scales.

The family is confined to North and Central America, where it is represented by five genera and 59 species. The greatest diversity occurs in Mexico and Guatemala, but two species, the Isthmian Alligator Lizard (*Gerrhonotus rhombifer*) and Montane Alligator Lizard (*Mesaspis monticola*), occur as far south as Costa Rica and Panama. The dominant genus in North America is *Elgaria*, with four US species, one of which, the Northern Alligator Lizard (*Elgaria coerulea*), ranges as far north as Canada, its northern range resulting in it being the only live-bearing species in an otherwise oviparous genus. The Mexican alligator lizards (*Gerrhonotus*), one of which, the Texas Alligator Lizard (*G. infernalis*), also reaches into the USA, are also oviparous, but the other three genera are viviparous.

Half of all alligator lizards are primarily terrestrial, but the southern Mexican to Central American genus *Abronia* contains almost 30 highly

DISTRIBUTION
North and Central America

GENERA
Abronia, Barisia, Elgaria, Gerrhonotus, and *Mesaspis*

HABITATS
Moist temperate woodland, coastal, montane, or tropical forests, and also savannas and deserts

SIZE
SVL 3 in (75 mm) Chiszar's Arboreal Alligator Lizard (*Abronia chiszari*) to 8 in (204 mm) Texas Alligator Lizard (*Gerrhonotus infernalis*)

ACTIVITY
Terrestrial, scansorial, or highly arboreal, living in epiphytes (*Abronia*); diurnal, but secretive

arboreal species, which may be found inhabiting bromeliads and other epiphytic plants at elevations up to 9,840 ft (3,000 m). Most alligator lizards are insectivorous, but a few also eat small vertebrates.

From 59 species of alligator lizards, ten species are listed as Vulnerable by the IUCN, 22 are listed as Endangered, while two species, Frost's Arboreal Alligator Lizard (*Abronia frosti*) and Campbell's Arboreal Alligator Lizard (*A. campbelli*), both from Guatemala, are listed as Critically Endangered.

REPRODUCTION
Most species are viviparous, giving birth to 1–12 neonates, but *Elgaria* contains both oviparous and viviparous species, and *Gerrhonotus* is oviparous, laying 5–20 parchment-shelled eggs

DIET
Primarily arthropods, including scorpions or centipedes, but some larger species eat other lizards, small mammals, birds' eggs, and fledglings

ABOVE LEFT | The Northern Alligator Lizard (*Elgaria coerulea*) ranges as far north as Canada. It is the only viviparous member of its genus.

ABOVE | The Mexican Green Arboreal Alligator Lizard (*Abronia graminea*) is an attractive species that inhabits bromeliads in cloud forest canopies. it is listed as Endangered by the IUCN.

ANGUIDAE—ANNIELLINAE
CALIFORNIAN LEGLESS LIZARDS

There are six species in the genus *Anniella*, and if it were not for the fact that they inhabit southern California and northwestern Baja California they could easily be mistaken for slow worms (*Anguis*, page 208), but until recently the Anniellinae was treated as a separate and distinct family, Anniellidae.

The Californian legless lizards are almost entirely fossorial, living in sand dunes. They spend the night deep down in the sand, avoiding the upper layers which are cooled by coastal fog, but with the morning sun they move upward into the warming sand. They may be found by turning over surface debris such as old boards or flat rocks.

Like slow worms, the Californian legless lizards are live-bearers, but while slow worms may produce up to 36 neonates, four is the maximum for the smaller Californian species. And while slow worms feed on earthworms and slugs, the Californian legless lizards feed on arthropods such as insects or spiders.

LEFT | The Northern Baja California Legless Lizard (*Anniella stebbinsi*) closely resembles a small European slow worm (*Anguis*).

ABOVE | Possessing movable eyelids but no ear openings, the California Legless Lizard (*Anniella pulchra*) also has similar traits to a slow worm.

DISTRIBUTION
Southwestern USA (California) and northwestern Mexico (Baja California)

GENUS
Anniella

HABITATS
Sandy woodland or scrub, to 5,090 ft (1,550 m)

SIZE
SVL 5¼ in (132 mm) Southern California Legless lizard (*Anniella stebbinsi*) to 7 in (180 mm) California Legless Lizard (*A. pulchra*)

ACTIVITY
Secretive and fossorial; diurnal

REPRODUCTION
Viviparous, giving birth to 1–4 neonates

DIET
Arthropods, including insects and spiders

XENOSAURIDAE
KNOB-SCALED LIZARDS

The knob-scaled lizards comprise a small family, containing a single genus of 12 species distributed across eastern and southern Mexico, and into Guatemala. They are characterized by the excessively knobbly appearance of their scales. They spend much of their lives in rocky crevices, but do not appear to be very social, crevices usually only containing a single lizard. It is believed they regulate their body temperature by thigmothermy, warming their bodies by contact with the rock. Diurnally active, but relatively lethargic, they feed on large invertebrates but will also eat small lizards. They do not practice caudal autotomy, but have an unusual defensive strategy, forcibly voiding the liquid contents of their bladders if handled. All known species are viviparous, producing up to eight neonates. The IUCN lists Newman's Knob-scaled Lizard (*Xenosaurus newmanorum*) and the Flat-headed Knob-scaled Lizard (*X. platyceps*) as Endangered, while the San Martin Knob-scaled Lizard (*X. grandis*) is listed as Vulnerable.

RIGHT | The Flat-headed Knob-scaled Lizard (*Xenosaurus platyceps*) exhibits the dorsoventrally flattened body of a crevice-dweller.

DISTRIBUTION
Eastern and southern Mexico, and Guatemala

GENUS
Xenosaurus

HABITATS
Low montane wet forest, cloud forest, and dry montane scrub, usually in rocky terrain

SIZE
SVL 4 in (103 mm) Zacualtipán Knob-scaled Lizard (*Xenosaurus tzacualtipantecus*) to 5¼ in (133 mm) San Martin Knob-scaled Lizard (*X. grandis*)

ACTIVITY
Terrestrial and saxicolous; diurnal

REPRODUCTION
All species are viviparous, producing litters of 2–8 neonates

DIET
Large arthropods and small lizards

DIPLOGLOSSIDAE
GALLIWASPS & SOUTH AMERICAN LEGLESS LIZARDS

The members of this family were at one time included in the alligator lizard subfamily (Gerrhonotinae, see page 210), and some authors still adopt that arrangement, but others treat them as a distinct but related family. There are three genera. Two of these contain lizards with well-developed limbs and elongate bodies and tails, known as the "galliwasps." The name may have evolved from the belief they were venomous, and the Jamaican Giant Galliwasp (*Celestus occiduus*) is thought to have been included in voodoo traditions. With their smooth, shiny-scaled scaly bodies the galliwasps resemble skinks of the genus *Mabuya* (page 134) much more than the alligator lizards. They are terrestrial in habit, although some species do occasionally climb.

The genus *Celestus* is found from Mexico to Panama, and also in the Greater Antilles (Hispaniola, Jamaica, Navassa, and the Cayman Islands), while *Diploglossus* ranges from Mexico to

DISTRIBUTION
Central and South America, and the West Indies

GENERA
Celestus, *Diploglossus*, and *Ophiodes*

HABITATS
Upland oak-pine–oak forest, over 3,280 ft (1,000 m) asl., but also lowland forest, grassland, and scrub

SIZE
SVL 2½ in (60 mm) Massif de la Selle Galliwasp (*Celestus macrotus*) to 13 in (330 mm) Striped Brazilian Legless Lizard (*Ophiodes striatus*)

ACTIVITY
Terrestrial or fossorial, occasionally arboreal; diurnal, crepuscular, or nocturnal

LEFT | Ingrid's Galliwasp (*Celestus ingridae*) is one of only two viviparous species in the genus *Celestus*.

ABOVE | The Rainbow Galliwasp (*Diploglossus monotropis*) from Costa Rica is a stunning lizard.

REPRODUCTION
Ophiodes and most *Celestus* are viviparous, with 2–5 neonates, while most *Gerrhonotus* are oviparous, laying up to 27 eggs

DIET
Arthropods, including insects and spiders

Ecuador, and in the Caribbean is found on Cuba, Puerto Rico, and Montserrat. The primary difference between the two genera relates to their claws. In *Diploglossus* the claw is entirely enclosed in a scaly sheath, but it is exposed below the sheath in *Celestus*. Most *Diploglossus* are oviparous and most *Celestus* are viviparous, but both contains species which use the opposite reproductive strategy. All diploglossid lizards feed on arthropods, such as insects and spiders.

The galliwasp with the strangest name is the Lost And Found Galliwasp (*C. laf*). The holotype was collected in the grounds of the Lost and Found Eco Hostel in Panama.

The third genus is *Ophiodes*, a viviparous genus containing six species of legless lizards that strongly resemble the European slow worms (*Anguis*, p 208). Fossorial rainforest inhabitants, they are found in Brazil, Bolivia, Paraguay, Uruguay, and Argentina.

HELODERMATIDAE
GILA MONSTER & BEADED LIZARDS

Prior to the discovery that monitor lizards (*Varanus*, pages 220-230) and the agamas and iguanas (Iguania, pages 166-205) possess venom glands in their mouths, the helodermid lizards were considered the only venomous lizards. There are currently five species of *Heloderma* and they are immediately recognisable by their thickset bodies, thick tails, and broad, powerful heads, covered in knobbly, bead-like scales.

The smallest species, the Gila Monster (*H. suspectum*), pronounced "heela," is found in southern California, Utah, Nevada, and New Mexico, and through most of Arizona, south into Sonora and northern Sinaloa, northwest Mexico. It is a black and pink or yellow inhabitant of semi-desert, but it also turns up in suburban gardens.

The subspecies of beaded lizard have been elevated to specific status. The northernmost is the largest, the Rio Fuerte Beaded Lizard (*H. exasperatum*), found in southern Sonora and Sinaloa. The Mexican Beaded Lizard (*H. horridum*) occupies the remainder of western Mexico. The Chiapas Beaded Lizard (*H. alvarezi*) is from Chiapas, Mexico, and just enters Guatemala, while the

LEFT | The Rio Fuerte Beaded Lizard (*Heloderma exasperatum*) is the northernmost beaded lizard. It may be terrestrial or arboreal but these lizards also spend a lot of time underground.

RIGHT | The Gila Monster (*Heloderma suspectum*) is the only heloderm within the US and also the smallest species with a shorter head than other heloderms.

DISTRIBUTION
Southwestern USA, western Mexico, and Guatemala

GENUS
Heloderma

HABITATS
Dry woodland and semidesert

SIZE
SVL 14¼ in (360 mm) Gila Monster (*Heloderma suspectum*) to 20½ in (520 mm) Rio Fuerte Beaded Lizard (*H. exasperatum*)

ACTIVITY
Terrestrial, subterranean, or occasionally arboreal (*H. exasperatum*); crepuscular or nocturnal

REPRODUCTION
All species are oviparous, laying up to 12 leathery-shelled eggs

DIET
Small mammals, birds, eggs, reptiles, and large invertebrates

Guatemalan Beaded Lizard (*H. charlesbogerti*) is endemic to the Motagua Valley in Guatemala. Beaded lizards are black or brown, with yellow, and they inhabit dry woodland and river valleys known as "arroyas."

Helodermids are venomous, but, unlike in snakes, their venom glands and the teeth that deliver the venomous bite are located in the lower jaws. Their venom is more for defense than prey capture, since they feed on defenseless nestling birds and rodents, and eggs. Bites are extremely painful and tenacious, but there have been no confirmed human fatalities, at least not since the dubious accounts of the 1930s.

SHINISAURIDAE
CHINESE CROCODILE LIZARD

The Shinisauridae contains a single species, the Chinese Crocodile Lizard (*Shinisaurus crocodilurus*), from southern China and northern Vietnam. This is a short-headed lizard with longitudinal rows of knobbly scales running down its body and tail. It is generally a drab brown or gray with reddish marks on the flanks, which help it blend into its surroundings.

The Chinese Crocodile Lizard inhabits relatively high-elevation evergreen forests (up to 4,920 ft/1,500 m), where it dwells along cold mountain streams, and it feeds primarily on invertebrates including snails, but also on fish and tadpoles. Because the nights at these elevations are cold, it adopts a diurnal, but secretive, lifestyle, basking on low branches to warm its body. It will dive into water to escape potential threats and remain submerged for up to 30 minutes.

This slow-moving lizard is listed as Endangered by the IUCN, being threatened by habitat loss and also potentially from poaching, as it is much sought-after in the pet trade.

BELOW | The Chinese Crocodile Lizard (*Shinisaurus crocodilurus*) inhabits a small area of southern China and northern Vietnam. It is considered Endangered due to habitat loss and poaching for the the pet trade.

DISTRIBUTION
Southern China and northeastern Vietnam

GENUS
Shinisaurus

HABITATS
Montane evergreen forest, in association with water

SIZE
SVL 11¾ in (300 mm) Chinese Crocodile Lizard (*Shinisaurus crocodilus*)

ACTIVITY
Semi-aquatic, and sometimes arboreal; diurnal and secretive

REPRODUCTION
Viviparous, producing 2–7 neonates

DIET
Crustaceans, snails, fish, and tadpoles

LANTHANOTIDAE
BORNEO EARLESS MONITOR

The Lanthanotidae also contains a single species, the Borneo Earless Monitor (*Lanthanotus borneensis*), a secretive, semi-fossorial inhabitant of lowland Bornean rainforest, where it forages for prey in wet leaf litter. It primarily feeds on invertebrates, such as earthworms and crustaceans, but being semi-aquatic it is also reported to eat fish. It is a drab brown lizard, to help it blend into the forest floor, with longitudinal rows of raised scales down its back and a slightly prehensile tail. It has small eyes with eyelids, a forked tongue like other monitor lizards (*Varanus*, page 220) and snakes, and, as its common name suggests, no external ear openings. Although believed to be closely related to the Varanidae (pages 220-230), it is not close enough to be included in that family.

The Borneo Earless Monitor is nocturnal and rarely seen, spending the day beneath rotten logs, in riverbank holes, or in deep leaf litter. It is known to occur around the Niah Caves in Sarawak, but its overall distribution in Borneo is not known.

RIGHT | The Borneo Earless Monitor Lizard (*Lanthanotus borneensis*) is a slow-moving, leaf-litter-dwelling lizard unique to the island of Borneo.

DISTRIBUTION
Western Borneo (Sarawak)

GENUS
Lanthanotus

HABITATS
Lowland rainforest, in association with water

SIZE
SVL 7¾ in (200 mm) Borneo Earless Monitor (*Lanthanotus borneensis*)

ACTIVITY
Semi-aquatic and semi-fossorial; nocturnal and secretive

REPRODUCTION
Oviparous, producing 2–5 oval, leathery-shelled eggs

DIET
Invertebrates such as earthworms and crustaceans, and fish

VARANIDAE—EMPAGUSIA & SOTEROSAURUS
ASIAN MONITOR LIZARDS

The Varanidae contains 84 species of monitor lizards distributed across Africa, Arabia, Asia, Australia, and Melanesia. The diversity of and relationships between these 84 species have led scientists to create 11 subgenera to contain them within the genus *Varanus*. If it were not for a desire to retain the genus *Varanus*, these subgenera could be raised to generic status.

The subgenus *Empagusia* contains five species of relatively large Asian monitor lizards (SVL 20¼–35½ in/515–900 mm) that inhabit lowland rainforest, dry forest, and plantations from Pakistan to Borneo and Java. South Asia is occupied by the Bengal Monitor (*V. bengalensis*), with the Yellow Monitor (*V. flavescens*) occurring in sympatry in the north and west of its range, while three species inhabit Southeast Asia, Dumeril's Monitor (*V. dumerili*), the Clouded Monitor (*V. nebulosus*), and the Rough-necked Monitor (*V. rudicollis*). Although primarily terrestrial, these lizards also swim and climb well.

The subgenus *Soterosaurus* contains ten primarily Southeast Asian species, but it is also represented in Bangladesh, northeast India, and Sri Lanka. Most of this range is occupied by the Common Asian Water Monitor (*V. salvator*), the largest specimens

DISTRIBUTION
Africa, Arabia, Asia, Australia, and Melanesia

GENUS
Varanus. Subgenera: *Empagusia, Euprepiosaurus, Hapturosaurus, Odatria, Papusaurus, Philippinosaurus, Polydaedalus, Psammosaurus, Solomonsaurus, Soterosaurus,* and *Varanus*

HABITATS
Tropical rainforest, dry woodland, grassland, desert and semidesert, wetlands and rivers, and islands

SIZE
SVL 4½ in (116 mm) Dampier Peninsula Goanna (*Varanus sparnus*) to 5 ft (1.54 m) Komodo Dragon (*V. komodoensis*)

ACTIVITY
Terrestrial, arboreal, saxicolous, fossorial, and arboreal; diurnal

(SVL 3¾ ft / 1.17 m) being from Sri Lanka. These lizards are highly aquatic and even oceangoing, being able to swim long distances, assisted by their laterally compressed and keeled tails, which serve as excellent rudders. Eight species inhabit the Philippines, including the Mindanao Water Monitor (*V. cumingi*) from Luzon, the Samar Water Monitor (*V. samarensis*), and the Palawan Water Monitor (*V. palawanensis*), while the species with the smallest range is the Togian Water Monitor (*V. togianus*) from the tiny Togian Islands east of Sulawesi, Indonesia.

ABOVE | The Komodo Dragon (*Varanus komodoensis*), a member of subgenus *Varanus* (page 230-1), is one of the most easily recognized Varanids and is the largest living lizard in the world.

BELOW | The Common Asian Water Monitor Lizard (*Varanus salvator*) is a large and very widely distributed species with five subspecies. A number of current species, such as the Mindanao Water Monitor (*V. cumingi*), are former subspecies.

REPRODUCTION
All species are oviparous, laying up to 42 leathery-shelled eggs

DIET
Primarily carnivorous, taking invertebrates, reptiles, birds, eggs, mammals, and carrion, but three species are omnivorous and also eat fruit

VARANIDAE — POLYDAEDALUS & PSAMMOSAURUS
AFRO-ARABIAN MONITOR LIZARDS

Sub-Saharan African is home to the subgenus *Polydaedalus*, which contains five species. The Nile Monitor (*Varanus niloticus*) occupies virtually this entire region apart from the most arid and southern areas, from South Africa to Senegal and Somalia, and along the Nile into Sudan and Egypt. The Ornate Monitor (*V. ornatus*) from the rainforests of West and Central Africa, is a former subspecies. The Nile Monitor is large (> 38½ in/980 mm), long-headed, long-tailed, boldly patterned, and both arboreal and aquatic. It feeds on almost anything it can swallow, and is a noted egg thief, raiding both crocodile and turtle nests.

The other three species in the subgenus are stockier, with shorter heads. They are also associated with more arid habitats than the Nile Monitor. The White-throated Monitor (*V. albigularis*) inhabits southern and eastern Africa, while the similar Bosc Monitor (*V. exanthematicus*) occurs in the Sahel, from Senegal to Somalia, between the Sahara and the rainforests. The rarest member of the subgenus is the Yemen Monitor (*V. yemenensis*), confined to Yemen and southwestern Saudi Arabia.

The subgenus *Psammosaurus* contains two species. Three subspecies of Desert Monitor (*V. griseus*) occur across North Africa and Arabia (*V. g. griseus*), Iran and the Caucasus (*V. g. caspius*), and the Pakistan–India border region (*V. g. koniecznyi*), while a second species, Nesterov's Desert Monitor (*V. nesterovi*), was recently described from the Iranian Zagros Mountains. All desert monitors are relatively slender, with long heads and tails. The Desert Monitor is sand-colored, with transverse dark bands across the back, while Nesterov's Desert Monitor lacks these bold bands.

LEFT | The Nile Monitor Lizard (*Varanus niloticus*) is widely distributed across sub-Saharan Africa. It climbs very well but it is also highly aquatic, the powerful tail bearing a dorsal keel to aid swimming.

ABOVE | The Desert Monitor Lizard (*Varanus griseus*) is the species in arid habitats across North Africa, the Arabian Peninsula and from Turkey to Iran and Afghanistan.

VARANIDAE—PHILIPPINOSAURUS, PAPUSAURUS & SOLOMONSAURUS
PHILIPPINE MONITOR LIZARDS

The subgenus *Philippinosaurus* contains three Philippine species. Gray's Monitor (*Varanus olivaceus*) inhabits southern Luzon, Polillo, and Catanduanes. It is a large (SVL 36½ in/925 mm), highly arboreal species. The most notable feature of its diet is that adults primarily eat fruit, showing a preference for the fruit of screw pines. When neither sugary nor oily fruits are available during the monsoon season, they switch to insects, arachnids, crustaceans, and mollusks. Until the start of this century, Gray's Monitor was thought to be the only frugivorous monitor lizard, but in recent years two more species have been described, the Panay Monitor (*V. mabitang*), and the Sierra Madre Monitor (*V. bitatawa*) from eastern Luzon.

LEFT | Salvador's Monitor Lizard (*Varanus salvadorii*) is a very large arboreal lizard found across the island of New Guinea, but despite its large size and wide distribution it is rarely encountered.

RIGHT | The Santa Isabel Monitor Lizard (*Varanus spinulosus*) is the sole member of the subgenus *Solomonsaurus*. This is a juvenile specimen.

The Papuan Monitor (*V. salvadorii*) is a Papuan endemic found throughout the entire island of New Guinea, except possibly the Central Cordillera, but it is not well known due to its secretive nature. It is the sole species in subgenus *Papusaurus*. This is a long species with a SVL of 33½ in (850 mm), but a TTL of up to 8¼ ft (2.5 m), due to its extremely long tail. The elongate, blunt head and yellow, dark-centered rosette pattern are distinctive. Agile and arboreal, it climbs well. Prey is thought to comprise primarily birds, but mammals, reptiles, and invertebrates are also likely to be taken. Villagers claim this lizard, which they call "Artrellia," also eats dogs.

The subgenus *Solomonsaurus* contains a single Solomon Islands endemic, the Santa Isabel Monitor (*V. spinulosus*), a relatively small species (SVL 12⅔ in/320 mm) from southeast Santa Isabel and neighboring San Jorge Island. It has a short head and a laterally compressed tail with a dorsal keel.

ABOVE | Gray's Monitor Lizard (*Varanus olivaceus*) was once thought to be the only omnivorous monitor lizard, but now two other species are known to adopt the same frugivorous diet. All three come from the Philippines.

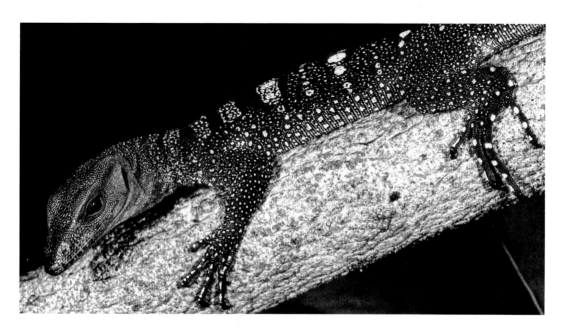

VARANIDAE—EUPREPIOSAURUS & HAPTUROSAURUS
MELANESIAN MONITOR LIZARDS

Apart from *Papusaurus* there are two other primarily Papuan subgenera. *Euprepiosaurus* contains 18 species of mangrove monitors. The most widely distributed species is the Mangrove Monitor (*Varanus indicus*), which occurs around the coast of New Guinea and the north coast of Australia. A black or bottle-green lizard covered with fine yellow spots, the Mangrove Monitor is arboreal and aquatic, and found in high numbers in coastal mangrove swamps, diving into the water and swimming away submerged to escape enemies.

Another widespread Papuan species is the striking Blue-tailed Monitor (*V. doreanus*), which inhabits New Guinea, the Bismarck Archipelago, and Bougainville to the east, and the Moluccas of Indonesia. This subgenus also contains some very remote island species, including the Talaud Mangrove Monitor (*V. lirungensis*) from the Talaud Islands, north of Sulawesi, the Mussau Monitor (*V. semotus*) from Mussau Island, north of New Ireland, and the Rennell Island Monitor (*V. juxtindicus*) from the Solomon Islands.

The monitors in *Euprepiosaurus* are primarily aquatic, but those in subgenus *Hapturosaurus* are highly arboreal rainforest species with slender bodies and prehensile tails, typified by the Emerald Tree Monitor (*V. prasinus*), from New Guinea, and the Black Tree Monitor (*V. beccarii*), from the Aru Islands. The nine species in the subgenus occur through the island of New Guinea and onto its satellite islands, such as the Rossel Island Monitor (*V. telenestes*) and Biak Monitor (*V. kordensis*), and Reisinger's Monitor (*V. reisingeri*), from Misool Island. Only one member of this subgenus occurs on the Australian mainland, the Nesbit River or Canopy Monitor (*V. keithhornei*) in eastern Queensland.

LEFT | The widely distributed Mangrove Monitor Lizard (*Varanus indicus*) is found in the coastal mangrove swamps of New Guinea, northern Australia, and eastern Indonesia.

RIGHT | The Emerald Tree Monitor Lizard (*Varanus prasinus*) is a highly arboreal New Guinea species that inhabits the rainforest canopy and possesses a long prehensile tail that helps it move around in the canopy

VARANIDAE—ODATRIA
AUSTRALIAN MONITOR LIZARDS

Most small to medium-sized Australian monitor lizards belong to the subgenus *Odatria*, which occurs across northeastern, northern, central, and western Australia. The largest species is the Black-palmed Rock Monitor (*Varanus glebopalma*, SVL 15⅔ in/397 mm), from the Top End of Australia, while the smallest is the Dampier Peninsula Monitor (*V. sparnus*, SVL 4½ in/116 mm), from Western Australia.

Most of the 22 species are terrestrial desert, arid grassland, or dry woodland inhabitants, typified by slender, long-tailed, sharp-nosed species such as the Pilbara Mulga Monitor (*V. bushi*) or Pygmy Mulga Monitor (*V. gilleni*), both of which inhabit mulga or gum trees. The Short-tailed Pygmy Monitor (*V. brevicauda*) is a semi-fossorial species inhabiting burrows in spinifex grassland, and is also the varanid with the shortest limbs.

Typical rock-dwellers include the Spiny-tailed Monitor (*V. acanthurus*), Kimberley Rock Monitor (*V. glauerti*), and Southern and Northern Pilbara Rock Monitors (*V. hamersleyensis* and *V. pilbarensis*). This subgenus also includes aquatic species such as Mitchell's Water Monitor (*V. mitchelli*), which inhabits rivers and lagoons in the Top End.

Not all species are Australian endemics. The Spotted Tree Monitor (*V. scalaris*), from northern Australia, also occurs in southern New Guinea, while two other species occur outside of Australia. The Timor Tree Monitor (*V. timorensis*) inhabits the dry coastal woodland on the island of Timor, while Auffenberg's Monitor (*V. auffenbergi*) is found on Roti Island, off the western coast of West Timor. In Australia monitor lizards are often called "goannas" —a derivation of iguana, a family of lizards to which they are not related.

BELOW | The Black-palmed Monitor Lizard (*Varanus glebopalma*) is a slender rock-dweller from northern Australia.

ABOVE | The Spiny-tailed Monitor Lizard (*Varanus acanthurus*) is a widely distributed rock-dwelling species from northwestern Australia that also climbs trees well.

BELOW | The Timor Tree Monitor Lizard (*Varanus timoriensis*) inhabits the coastal forests and woodlands of arid northern and wet southern Timor.

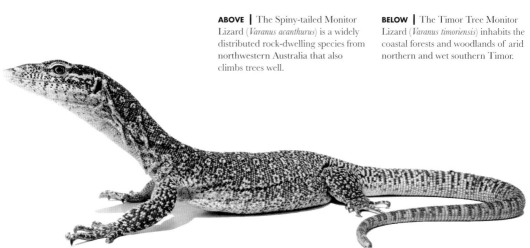

VARANIDAE—VARANUS
INDO-AUSTRALIAN MONITOR LIZARDS

The nominate subgenus, *Varanus*, contains the largest lizard species in the world, the Komodo Dragon (*V. komodoensis*, SVL 5 ft/1.54 m), which inhabits a small group of islands in the Lesser Sundas of Indonesia, comprising Komodo, Rinca, Gili Motang, and western Flores. They no longer live on Padar. This large lizard stalks prey such as deer, goats, wild boar, or water buffalo, rushing forward to deliver a bite and then following the prey until it succumbs to the venom. That monitor lizards are venomous is a relatively recent discovery; previously, Komodo Dragons were thought to subdue prey through infection caused by bacteria in their saliva This ambush tactic may have resulted from when they fed on more dangerous prey, an extinct dwarf subspecies of elephant (*Stegodon florensis insularis*) that occurred in the region during the Pleistocene. There have also been rare incidents where Komodo Dragons have killed humans. However, the Komodo Dragon is not he largest lizard to have ever existed. Megalania (*V. priscus*) was a huge species (18–26 ft/ 5.5–7.9 m TTL) that existed in Australia when humans first arrived 65–40,000 years ago.

Varanus also includes Australia's largest living lizards, the Perentie (*V. giganteus*, SVL 37 1/3 in/ 950 m), from central and western Australia, and the Lace Monitor (*V. varius*, SVL 30 in/765 mm), from eastern Australia. The smallest species is the Heath Monitor (*V. rosenbergi*, SVL 20 1/3 in/518 mm), which inhabits heathlands and wet forests along Australia's southern coast.

While most members of subgenus *Varanus* are terrestrial, one species is highly aquatic, Mertens' Water Monitor (*V. mertensi*), which inhabits coastal brackish and freshwater rivers and lagoons in the Top End.

The Argus Monitor (*V. panoptes*) is the most widely distributed species, occurring across much of western and northern Australia, with one subspecies (*V. p. horni*) inhabiting the southern savannas of New Guinea.

LEFT | Merten's Water Monitor Lizard (*Varanus mertensi*) is a freshwater aquatic species from northern Australia.

ABOVE | The Komodo Dragon (*Varanus komodoensis*) might be the "King of Lizards" today, but it would have looked small alongside the Megalania (*V. priscus*) from Pleistocene Australia.

GLOSSARY

Arboreal: living in the trees (see also terrestrial, fossorial).

Autotomy: deliberately discarding a body part (tail = caudal autotomy, skin = dermal autotomy) to avoid predation.

Basal: close to the evolutionary origin of the group.

Bipedal: running on the two back legs.

Casque: a raised ornamental protuberance on the head.

Chaparral: a North American habitat comprising thorny bushes and shrubs.

Clade: any taxonomic group containing an ancestor and all its descendants but not necessarily having a taxonomic hierarchical status, such as family or genus.

Cloaca: the common genital-excretory opening of reptiles.

Congener, congeneric: a member of the same genus.

Convergent evolution: (also known as parallel evolution) when two unrelated species in different places evolve to look similar because they occupy similar niches.

Crepuscular: active at dusk or dawn (see also diurnal, nocturnal).

Cryptic: camouflaged and difficult to see.

Dewlap: a flap of skin under the throat of some lizards, usually iguanine or agamid lizards, that can be used for signaling to attract a female or defend territory.

Diurnal: active by day (see also crepuscular, nocturnal).

Dorsal, dorsum: upper surface (see also ventral, lateral).

Dorsolateral: on the upper sides.

Dorsoventrally compressed: a body shape where the lizard is flattened to produce a large surface area to the sun for basking.

Ectothermic: raising the body temperature by basking in the sun or by radiation from already warmed surfaces, also known as cold-blooded (see also endothermic).

Endemic/endemism/endemicity: an organism found in a single location, such as an island or a mountain above a certain elevation, which is therefore vulnerable to changes due to humans or the climate.

Endothermic: maintaining a constant body temperature metabolically, also known as warm-blooded (see also ectothermic).

Estivate: to remain dormant in hiding, through a period of extreme aridity or high temperatures, such as the tropical dry season.

Fossorial, semi-fossorial: to live underground as a burrower, or in the sub-soil as a sometimes burrower.

Fynbos: a South African habitat comprising low-growing shrubland/heathland plants, often in coastal areas.

Genus (pl. genera): the first part of a binomial scientific name, a genus contains species that are more related to each other than they are to species in another genus (see also species).

Heliophilic: diurnal organisms that actively seek out and bask in the sun.

Heliophobic: organisms that avoid daylight, preferring shade or darkness, i.e. many fossorial or nocturnal species.

Hemipene: one of the paired male sexual organs of lizards and snakes.

Herpetofauna: the term used to encompass all the reptiles and amphibians of a particular region, country, or habitat.

Holotype: the single museum specimen used in the description of a new species, the official name-bearer for that species with which other specimens are compared. Holotypes are extremely important and biologically valuable so museums often store them in fire-proof cabinets.

Igapó forest: a Brazilian habitat in the low reaches of rivers or around the periphery of lakes comprising seasonally blackwater-flooded forests that experience vast differences in water level every year, from high and dry to several yards of submergence.

Karoo: a South African habitat comprising semi-desert.

Keel/keeled: a ridge down the longitudinal center of each scale that gives a lizard a rough appearance and texture, sometimes keeled or carinate scales have several such ridges.

Lamellae (sing. lamella): in lizards (e.g. geckos, anoles) the rows of thin, plate-like structures on the underside of the dilated toe that contain millions of tiny adhesive setae that enable the lizards to grip and run up smooth surfaces.

Lateral: of the sides of the body (see also dorsal, ventral).

Laterally compressed: a body form that is compressed to be narrow and tall, such as chameleons.

Melanistic: containing high levels of melanin, the pigment in the skin that makes an organism dark; an advantage in cold climates where dark surfaces warm in the sun quicker than light surfaces.

Mesophytic – thriving in habitats with moderate water supply and temperatures.

Monophyletic: a group that contains a common ancestor and all the organisms that have descended from that same common ancestor and form a natural evolutionary clade, i.e. Lepidosauria which contains the Tuatara and all lizards, worm-lizards, and snakes and their common ancestor (see also paraphyletic).

Monotypic: a family containing a single genus or a genus containing a single species.

Multi-clutching: producing more than one clutch of eggs in a single year/season.

MYA: million years ago.

Neonate: a newborn offspring of a viviparous species.

Nocturnal: active at night (see also diurnal, crepuscular).

Nominate genus: the genus that provides the family in which it is contained with its name, e.g. *Iguana* in Iguanidae.

Oviparous: species in which the females lay eggs (hard or soft shelled) as a means of reproduction.

Paraphyletic: a group that contains an ancestor and some, but not all, of its descendants, i.e. living Reptilia without Aves (birds) included, or Sauria without snakes (see also monophyletic).

Parthenogenetic: virgin birth, females that reproduce and produce viable offspring without the requirement of mating with a male. Obligate parthenogens (such as some Pacific geckos and Middle American whiptail lizards) always reproduce this way. Facultative parthenogens are usually sexual species adopting this method of reproduction in the absence of any males, such as the Komodo dragon (*Varanus komodoensis*).

Patagium (pl. patagia): a gliding organ comprising an abundance of loose skin stretched over elongate ribs that enables a flying lizard (*Draco*) to leap from a tree and escape to safety.

Pentadactyl: five-fingered (or toed).

Perianthropic: living alongside humans or in human-mediated habitats.

Phylogeny: the evolutionary history of a species or group of species, often illustrated as a tree-like diagram constructed to demonstrate the inter-relatedness of species, genera, and higher taxa, and the route back to their common ancestor.

Pineal eye: a third eye located on the dorsum of the head, which may or may not be visible as a small opaque lens in the center of the head, and which is used for thermoregulation.

Prehensile: able to grasp, usually in relation to a tail which can be used as an additional limb for climbing.

Rugose: rough skinned or scales, highly keeled.

Saxicolous: rock dwelling.

Scansorial: able to climb.

Sexual dichromatism: when the sexes are different colors or patterns, such as breeding colors in males.

Sexual dimorphism: when the sexes look physically different because of morphological characteristics ranging from large heads in males to the presence of ornate crests, casques, or other adornments, although the differences may be subtle.

Sister taxon: the most closely related group or species to the one being discussed, i.e. two closely related species in a phylogeny may be termed sister-taxa.

Speciate, speciation: the process by which populations evolve to produce new species.

Species: the basic biological unit, a group of biologically related organisms that can reproduce together and produce viable offspring. However, what constitutes a species is open to various interpretations. Sub-units within a species are subspecies.

Species complex: a group of closely related species which are morphologically difficult to distinguish apart.

Specific status: if a subspecies of a species is considered sufficiently distinct from the other subspecies in that species it may be elevated to a species itself, often due to the molecular analysis of its DNA in relation to the DNA of the other subspecies.

Spectacle/brille: a transparent covering over the eye that enables the organism to see out but which protects the eye against damage, and which is sloughed during ecdysis (skin shedding); found in all snakes and some lizards (most geckos, some skinks).

Squamate: a member of the order Squamata (scaled reptiles) that contains all snakes, lizards, and worm-lizards, the sister-taxa to the Rhynchocephalia which contains only one living species, the tuatara (*Sphenodon punctatus*).

Subgenus (pl. subgenera): a hierarchical level between genus and species. It is usually only used in large genera to link groups of related species together as distinct from other groups of species within the same genus, such as in the monitor lizards (*Varanus*), with the subgenus usually cited in parentheses, e.g. *Varanus (Papusaurus) salvadorii*.

Sympatry: when two species occur side-by-side in the same biological niche or habitat.

Taxon (p. taxa): any hierarchical taxonomic unit, i.e. phylum, family, genus, species, a clade with hierarchical status.

Tepui: a steep-sided, table-topped mountain found in Venezuela or Brazil, which often exhibits a high degree of endemicity on its flat top due to the flora and fauna's long-term isolation from the lowland rainforest surrounding it.

Terrestrial: living on the ground.

Tetrapod: a four-limbed vertebrate; all vertebrates above the level of the fishes which illustrate an evolutionary history back to the first amphibians that moved onto land. Snakes and many lizards lack limbs but they are still tetrapods because their ancestors had four limbs.

Ventral, venter: of the undersides of the body (see also dorsal, lateral).

Vestigial: the remains of a characteristic that has otherwise been lost through evolution. The coccyx in humans is a vestigial tail; limbs are vestigial in some elongate fossorial lizards.

Viviparous: species in which the females produce live young as a means of reproduction.

Xerophytic: adapted for life in dry habitats where water is scarce.

RESOURCES

BOOKS (GENERAL)

Bauer, A. M. 2013
Geckos: The Animal Answer Guide. John Hopkins University Press.

Burghardt, G. M. & A. S. Rand 1982
Iguanas of the World: Their Behavior, Ecology, and Conservation. Noyes Publications.

King, D., E. R. Pianka & R. A. King 2004
Varanoid Lizards of the World. Indiana University Press.

Necas, P. 1999
Chameleons: Nature's Hidden Jewels. Edition Chimaira.

Pianka, E. R. & L. J. Vitt 2006
Lizards: Windows to the Evolution of Diversity. University of California Press.

Pough, F. H., R. M. Andrews, M. L. Crump, A. H. Savitsky, K. D. Wells & M. C. Bradley 2016
Herpetology (4th edition). Sinauer Publishing.

Rodda, G. H. 2020
Lizards of the World: Natural History and Taxon Accounts. John Hopkins University Press.

Vitt, L. J. & J. P. Caldwell 2014
Herpetology: An Introductory Biology of Amphibians and Reptiles (4th edition). Academic Press.

FIELD GUIDES
(NATIONAL & STATE GUIDES NOT LISTED)

NORTH AMERICA

Powell, R. & R. Conant 2016
Peterson Field Guide to the Reptiles and Amphibians of Eastern and Central North America (4th edition). Houghton Mifflin Harcourt.

Stebbins, R. C. & S. M. McGinnis 2018
Peterson Field Guide to the Reptiles and Amphibians of Western North America (4th edition). Houghton Mifflin Harcourt.

CENTRAL AND SOUTH AMERICA, AND WEST INDIES

Arteaga, A., L. Bustamante; J. Vieria; W. Tapia; J. M. Guayasamin 2019
Reptiles of the Galápagos: Life in the Enchanted Islands. Tropical Herping.

Bartlett, R. D. & P. P. Bartlett 2002
Reptiles and Amphibians of the Amazon: An Ecotourist's Guide. University of Florida Press.

Bezy, R. L. 2019
Night Lizards: Field Memoirs and a Summary of the Xantusidae. ECO Publishing.

Crother, B. I. 1999
Caribbean Amphibians and Reptiles. Academic Press.

Köhler, G. & L. D. Wilson 2003
Reptiles of Central America. Herpeton Verlag

Lemm, J. & S. C. Alberts 2012
Cyclura: Natural History, Husbandry, and Conservation of West Indian Rock Iguanas. Academic Press.

Savage, J. M. 2002
The Amphibians and Reptiles of Costa Rica. University of Chicago Press.

Schwartz, A. & R. W. Henderson 1991
Amphibians and Reptiles of the West Indies: Descriptions, Distributions, and Natural History. University of Florida Press.

EUROPE

Arnold, E. N. & D. W. Ovenden 2004
Reptiles and Amphibians of Europe. Harper Collins.

Beebee, T. & R. Griffiths 2009
Amphibians and Reptiles of Europe. HarperCollins.

Speybroeck, J., W. Beukema, B. Bok, J. van der Voort & I. Velikov 2016
Field Guide to the Reptiles and Amphibians of Britain and Europe. Bloomsbury.

AFRICA & MADAGASCAR

Branch, B. 1998
Field Guide to Snakes and Other Reptiles of Southern Africa. Struik.

Glaw, F. & M. Vences 1994
A Field Guide to the Amphibians and Reptiles of Madagascar (3rd edition). Vences & Glaw Verlag.

Henkel, F-W. & W. Schmidt 2000
The Amphibians and Reptiles of Madagascar, the Mascarenes, the Seychelles and the Comoros Islands. Krieger Publishing.

Howell, K., S. Spawls, H. Hinkel & M. Menegon 2017
Field Guide to East African Reptiles (2nd edition). Bloomsbury.

Necas, P. & W. Schmidt 2004
Stump-tailed Chameleons: Miniature dragons of the rainforest. Edition Chimaira.

Reissig, J. 2014
Girdled Lizards and their relatives. Edition Chimaira.

Tilbury, C. 2010
Chameleons of Africa: An atlas including the chameleons of Europe, the Middle East and Asia. Edition Chimaira.

ASIA & ARABIA

Chan-Ard, T., J. W. R. Parr & J. Nabhitabhata 2015
A Field Guide to the Reptiles of Thailand. Oxford University Press.

Das, I. 2015
A Field Guide to the Reptiles of South-East Asia. Bloomsbury.

AUSTRALASIA & OCEANIA

Cogger H. G. 2014
Reptiles and Amphibians of Australia (7th edition). CSIRO Publishing.

McCoy, M. 2006 *Reptiles of the Solomon Islands*. Pensoft.

Swan, S. K. & G. Swan 2017
A Complete Guide to the Reptiles of Australia (5th edition). Reed/New Holland.

HERPETOLOGICAL SOCIETIES

There are also numerous national or statewide societies

Society for the Study of Reptiles and Amphibians (SSAR) *ssarherps.org*

American Society of Ichthyologists and Herpetologists (ASIH) *www.asih.org*

Herpetologists' League (HL) *herpetologistsleague.org*

Societas Europaea Herpetologica (SEH) *www.seh-herpetology.org*

Australian Herpetological Society (AHS) *www.ahs.org.au*

Herpetological Association of Africa (HAA) *www.africanherpetology.org*

British Herpetological Society (BHS) *www.thebhs.org*

Deutsche Gesellschaft für Herpetologie und Terrarienkunde (DGHT) *www.dght.de/startseite*

Société Herpétologique de France (SHF) *lashf.org*
Societas Herpetologica Italica (SHI)
www-9.unipv.it/webshi/societa/societas.htm

International Herpetological Society (IHS) *www.ihs-web.org.uk*

USEFUL WEBSITES

World Congress of Herpetology (WCH)
www.worldcongressofherpetology.org

Reptile Database: *reptile-database.reptarium.cz*
(use Advanced Search facility)

International Herpetological Symposium (IHS)
www.internationalherpetologicalsymposium.com

International Union for the Conservation of Nature (IUCN) Red List of Threatened Species *www.iucnredlist.org*

Convention on International Trade in Endangered Species of Fauna and Flora (CITES) *www.cites.org*

World Association of Zoos and Aquariums (WAZA)
www.waza.org/en/site/home

INDEX

A

Abronia 210–11
Acanthodactylus 40, 157
Acontias 47, 127
Acontinae 21, 119, 126–7
Acrodonta 24, 167
Aeluroscalabotinae 106
Agama 50, 168–9
agamas 20, 29, 31, 35, 36, 41, 45, 50, 168–9
Agamidae 59, 167, 168–77
Agaminae 168–9
Agamodon 164
Ahaetulla 23
ajolotes 22, 46, 161
alligator lizards 207, 210–11
Alopoglossidae 141, 146–7
Alopoglossus 146
Amblyrhynchus 39, 58, 190, 203
Amphibolurinae 170–1
Amphisbaena 64–5, 158
Amphisbaenia 12, 17, 22, 25, 35, 46, 64–5, 141, 158–9
Amphisbaenidae 141, 158–9
anatomy 18–27
Anelytropsis 89
Anguidae 208–11
Anguimorpha 12, 21, 206–31
Anguinae 207, 208–9
Anguis 20, 21, 30, 33, 73, 208–9
Anisolepis 196
Anniella 21, 212
Anniellinae 207, 212
anoles 43, 49, 52, 54, 70, 196–7, 200–1
Anolis 49, 52, 54, 70, 72, 200–1
Antsingy Leaf Chameleon 183
Aprasia 35, 98, 99
aquatic lizards 41
Arabian Sandfish 47, 73
Archosauria 17
Argus Monitor 230
arid environments 38
Aristelliger 114
Armadillo Lizard 77, 120–1
Aruba Whiptail 60
Asaccus 111
Austral geckos 91, 94–7
Australian earless dragons 35
Australo-Asian skinks 21, 132–3
Austral skinks 119, 136–7
autotomy 6, 73–4, 208

B

Bachia 21, 22, 52, 150
banded geckos 76, 108
Basiliscus 41, 72, 73, 171, 195
basilisks 41, 72, 73, 171, 195
basking 18, 36, 50
beaded lizards 29, 30, 73, 207, 216–17

Bearded Dragon 59, 171
bent-toed geckos 42, 104, 114
Bipedidae 141, 161
Bipes 22, 46, 161
Blanidae 141, 160
Blanus 160
blind-lizards 20, 89
Blue Iguana Recovery Program 83–4, 186
blue-tailed skinks 128, 137
blue-tongued skinks 31, 56, 60, 70, 130
Borneo Earless Monitor 34, 35, 207, 219
Brachylophus 58, 191
Brasiliscincus 56, 134
Breyer's Long-tailed Seps 20
broad-tailed geckos 19, 93
Brookesia 183
Bufoniceps 41, 169
burrowing *see* fossorial lifestyle
burrowing skinks 21, 128–9
Burton's Snake-lizard 23, 66, 98
bushfires 79
Bush Para-anoles 199
Bushveld Lizard 76
butterfly lizards 18, 59, 174–5

C

Cachryx 185
Cadea 162
Cadeidae 141, 162
caiman lizards 25, 48, 64, 144–5
Californian legless lizards 21, 207
Callopistes 142
Callopistinae 142
Calotes 172
Calumma 182
Calyptommatus 47, 149
camouflage 70, 106
Canary Island lizards 152–3
cannibalism 67, 69
Cape Flat Lizard 60
carnivores 63–7
Carphodactylidae 91, 92
Carphodactylus 93
casque-headed lizards 195
Cat Gecko 108
Celestus 214–15
Centralian Knob-tailed Gecko 19
Cercosaura 150
Cercosaurinae 150–1
Chalarodon 205
Chamaeleo 48, 66, 86, 179–81
Chamaeleonidae 167, 178–83
Chamaesaura 20, 22, 121
chameleon geckos 43, 74, 93, 96–7
chameleons 33, 43, 48, 50, 65, 66, 70, 178–83
Chirindia 52

Chlamydosaurus 40, 41, 70, 72, 73, 171
Chondrodactylus 102–3
Christinus 105
Christmas Island 82–3
chuckwallas 40, 58, 189
clades 14
climbing 42–3
cloaca, transverse 6, 17
Cnemidophorus 48, 60, 142
Coeranoscincus 133
cold-climate hypothesis 55
cold environments 36
Coleonyx 76, 108
collared lizards 69, 194
Conolophus 58, 191
conservation 78–85
Cophosaurus 35, 202
Cordylidae 20, 77, 119, 120–1
Cordylinae 120–1
Cordylosaurus 122
Cordylus 120
Correlophus 96
Corucia 43, 56, 60, 74, 130
Corytophanes 42, 195
cosmopolitan geckos 91, 100–107
courtship 49–50, 51
cranial movement 23
Crested Gecko 96
Cricosaura 124
Cricosaurinae 124
croaking geckos 114
crocodile lizard 207, 218
crocodile skinks 52, 131
Crocodile Tegu 144–5
Crocodilurus 144–5
Crotaphytidae 194
Crotaphytus 69, 194
crypsis 70
Cryptagama 70
Cryptoblepharus 136
Ctenoblepharys 198
Ctenophorus 51, 170
Ctenosaura 42, 58, 185
Ctenotus 132
Cuban Night Lizard 124
Cuban worm-lizards 162
Curaçao Whiptail 60
curlytails 192
Cyclura 58, 83–4, 186
Cyrtodactylus 42, 104
Cyrtopodian 42

D

dabb lizards 40, 59, 73, 176–7
Dactyloidae 43, 200–1
Darevskia 57
Darwin's Marked Gecko 54
day geckos 33, 60, 79, 81, 102, 104, 106, 116

Delma 99
Dendrosauridion 150–1
dentition 24–5
dermal autotomy 74
desert iguanas 29, 33, 58, 188–9
Desert Plated Lizard 60
dewlap coloration 49, 50
dewlaps 173
Dibamia 12, 32
Dibamidae 89
Dibamus 20, 89
Dicrodon 60
diet 58–67
Diplodactylidae 91, 94–7
Diploglossidae 207, 214–15
Diploglossus 214–15
Diplolaemus 196
Dipsosaurus 29, 33, 58, 188–9
display 18, 49–50, 173
Dopasia 21
Dracaena 25, 48, 64, 144–5
Draco 18, 45, 49, 63, 70, 73, 172
Draconinae 172–3
dragon lizards 120
dwarf geckos 91, 103, 114–17

E

ears 34–5, 46
Eastern Blue-tongued Skink 56
Eastern Four-fingered Skink 7
ecdysis 28
Egernia 56, 60, 73, 131
Egerniinae 119, 130–1
Elgaria 210
Emoia 136–7
Enyaliinae 196–7
Enyalioides 193
Enyalius 196
Eremiadinae 156–7
Eremias 156–7
Eublepharidae 91, 108–9
Eublepharinae 108
Eublepharis 108
Eugongylinae 119, 136–7
Eugongylus 137
Euleptes 116, 117
Eurasian lizards 154–5
Eurydactylodes 43, 74, 97
Eutropis 54, 134
evolution 8–13, 48, 78
extinction 78–9, 82–5
eyelid geckos 91, 108–9
eyes 32–4, 46

F

Fan-throated Lizards 50, 173
femoral pores 29
feral species 80–1
Feylinia 47
fighting 50, 70–3
Fijian iguanas 58
fish-scaled geckos 106
Five-fingered Ajolote 46, 161
Five-lined Skink 54
Flannery, Tim 79
flap-footed lizards 98–9
flat lizards 121
flat-tailed geckos 70, 106
flick-leapers 99
Florida worm-lizards 163
flying dragons 45, 49, 63, 70, 73, 172
flying geckos 44, 73, 105
Fojia Mountain Skink 132
fossorial lifestyle 6, 12, 15, 22, 35, 46–7
Frilled Lizard 41, 70, 73, 171
fringe-toed lizards 40–1, 42
Furcifer 43, 65, 179, 182

G

Galapagos land iguanas 58, 190–1
Galapagos Marine Iguana 39, 58, 190, 203
galliwasps 76, 207, 214–15
Gallotia 60, 152–3
Gallotinae 60, 152–3
Gambelia 194
garden lizards 172
Gargoyle Gecko 59
Garthia 113
Geckolepis 106
geckos 19, 20, 29, 31, 33, 41, 42–3, 44–5, 52, 54, 57, 59–60, 69, 70, 73, 74, 76, 79, 81, 83, 90–117
Gekkonidae 91, 100–107
Gekkota 12, 20, 59–60, 73–4, 90–117
Gerrhonotinae 207, 210–11
Gerrhonotus 210
Gerrhosauridae 20, 119, 122–3
Gerrhosaurinae 122
Gerrhosaurus 41, 60, 122
Giant Ground Gecko 103
Gila Monsters 29, 30, 207, 216–17
girdled lizards 20, 77, 119, 120–1
glass lizards 21, 207, 208–9
gliding 44–5, 73
glottis 26
Gonatodes 114
Gongylomorphus 128–9
Goniurosaurus 108
Graceful Crocodile Skink 52, 131
grass lizards 20, 22, 121
Gray's Monitor 60, 224
Greater Earless Lizard 35
green-blooded skinks 74–5
green iguanas 42, 50, 58, 59, 84, 184–5
ground skinks 134–5
Guianan Caiman Lizard 25, 48, 62, 64
Gymnodactylus 113
Gymnophthalmidae 21, 52, 141, 148–51
Gymnophthalminae 148–9
Gymnophthalmus 148–9

H

Haacke-Greer's Skink 139
Haemodracon 111
hatching 54
heart 26
Heath's Skink 56
Heliobolus 76
Helmeted Gecko 33
helmeted lizards 42
Heloderma 29, 30, 73, 216–17
Helodermatidae 207, 216–17
Hemicordylus 120
Hemidactylus 57, 69, 83, 98–9, 106
hemipenes 6, 51
Hemiphyllodactylus 57, 101
herbivory 12, 25, 58–60
Holbrookia 35, 202
Homonota 54, 113
Hoplocercidae 193
Hoplocercus 193
Hoplodactylus 81–2, 96
horned lizards 30, 38, 54, 63, 70, 74, 172, 202–3
hot environments 37
house geckos 57, 69, 83, 100
human impact 78–83
Hyalosaurus 21, 209
Hydrosaurinae 174–5
Hydrosaurus 41, 59, 174

I

Iguana 42, 50, 58–9, 74, 84, 184–5, 186
iguanas 29, 31, 33, 34, 39, 42, 50, 54, 58, 59, 72, 83–4, 184–5, 186, 188–9, 190–1, 203
Iguania 12, 166–205
Iguanidae 58–9, 167, 184–91
insectivores 60–3
Intellagama 170
invasive species 80–1
Isabel Monitor 224

J

Jackson's Three-horned Chameleon 50
Jarujinia 22

K

Kawekaweau 81–2, 96
keel-scaled teiids 42
Kentropyx 42
keratin 28
keystone species 79
knob-scaled lizards 207, 213
knob-tailed geckos 92, 93
Komodo Dragon 11, 51, 57, 66–7, 230
Kudnu mackinleyi 9

L

Lace Monitor 11, 230
Lacerta 15, 50, 55, 154–5

Lacertidae 40, 60, 141, 152–7
Lacertilia 17, 21, 22
Lacertinae 154–5
Lacertoidea 12, 74, 140–65
Laemanctus 195
Lamprolepis 118–19, 139
Lampropholis 137
lance skinks 21, 47, 119, 126–7
Lanthanotidae 207, 219
Lanthanotus 34, 35, 219
Laudakia 36
lava lizards 204
Leach's Giant Gecko 19, 96
leaf-tailed geckos 93
leaf-toed geckos 91, 101, 102, 110–13, 116
lecithotrophy 52
legless lizards 21, 207, 212–14
leglessness 12, 20–2, 46
Leiocephalidae 192
Leiocephalus 192
Leiolepidinae 167, 175
Leiolepis 18, 57, 59, 175
Leiosauridae 196–7
Leiosaurinae 196–7
leopard geckos 108
leopard lizards 194
Lepidodactylus 101
Lepidophyma 60, 124–5
Lepidophyminae 124–5
Lepidosauria 8, 17
Lerista 21, 132
Lesser Antillean Iguana 186
Lesser Caymans Iguana 84
lesser earless lizards 35
Lialis 23, 66, 98, 99
Lialisinae 98
lingual prehension 31
Liolaemidae 59, 198
Liolaemus 36, 59, 198
litter size 56
lizard-fingered geckos 116
locomotion 19, 40–7
Long-tailed Grass Lizard 19
Long-tailed Sun Skink 54
lungs 26
Lygodactylus 103
Lygosominae 119, 138–9

M

Mabuyinae 119, 134–5
Malagasy Swift 205
Mallee Sand Dragon 51
Mangrove Monitor 226
manticores 193
marked geckos 113
mastigures 59, 73, 176–7
matrotrophy 56
Mauritius Blue-tailed Day Gecko 60, 79
Mbanja Worm-lizard 52
Megachirella wachtleri 8
Megalania 11, 67, 80, 230
melanism 36

Melanoseps 47
Meroles 37, 38, 157
Mesaspis 210
microteiids 21, 33, 47, 57, 141, 148–51
mimicry 76
Mochlus 138
molluskophagous diet 63–4
Moloch 38, 63, 171
monitor lizards 11, 19, 30, 34, 35, 50, 54, 60, 64, 66–7, 72, 207, 219–31
Monkey-tailed Skink 56, 60, 74, 130
Mop-headed Lizard 41, 204
Morunasaurus 193
myrmecophagy 63

N

Namaqua Chameleon 181
Namibian Desert Plated Lizard 41
nasal organs 30
Naultinus 96
Negative Pressure Ventilation 26
Nephrurus 19, 92
nests 54
night lizards 119, 124–5
Nile Monitor 54, 64, 222

O

Oberhaütchen 28
Ophidiocephalus 99
Ophiodes 21, 215
Ophisaurus 21, 209
Ophisops 157
Opluridae 205
Oplurus 205
osteoderms 28, 29, 77
Ouroborus 77, 121
Oustalet's Chameleon 65, 182
oviparity 52–4
Oxybelis 23

P

Paliguana whitei 9
Panther Chameleon 43, 182
Papuan Green-blooded Skink 43
Papuan Monitor 19, 224
Paradelma 99
parasites 68
Parson's Chameleon 182
parthenogenesis 57
patagium 18, 45, 172
Perentie 11, 230
Persian Wonder Gecko 41
Phelsuma 33, 60, 79, 106
Philippinosaurus 60
Phrynosoma 30, 38, 54, 56, 63, 70, 74, 202
Phrynosomatidae 40, 202–3
Phrynosomatinae 202–3
Phyllodactylidae 91, 110–13
Phyllodactylus 112
Phyllopezus 113
Phyllurus 19, 93
Phymaturus 59, 198

Physignathus 170
pineal eye 34
plated lizards 20, 119, 122–3
plate tectonics 10–11
Platysaurinae 121
Platysaurus 60, 121
Plestiodon 54, 128
Pletholax 99
Pleurodonta 24, 167
Plica 63, 70
Plumed Basilisk 41
Pogona 59, 171
Polychrotidae 199
Polychrus 199
Prasinohaema 43, 74–5, 132–3
predators of lizards 68–77
Pristidactylus 197
Pristurus 76, 116
Psammodromus 153
Pseudemoia 56
Pseudopus 21, 30, 33, 48, 64, 73, 208
Pseudothecadactylus 95
Ptychoglossus 146
Ptychozoon 44, 73, 105
Ptyodactylus 111
pygmy chameleons 70, 180–1
Pygmy Short-horned Lizard 56
Pygopodidae 91, 98–9
Pygopodinae 98
Pygopus 99

Q

Quedenfeldtia 116

R

racerunners 142
Rainbow Whiptail 48, 60
Rajasthan Toad-headed Agama 41, 169
Raukawa Gecko 59
reproduction 48–57
respiration 26
Réunion Skink 128–9
Rhachisaurinae 149
Rhacodactylus 19, 59, 96
Rhampholeon 70, 180
Rhineura 163
Rhineuridae 141, 163
Rhynchocephalia 8, 10, 17, 88
Riolaminae 149
Rodrigues Island 79
Round Island 81
running 19, 37, 40–1

S

Saara 59, 73, 176
sailfin lizards 41, 59, 174–5
saline environments 38–9
salt excretory glands 39
Saltuarius 93
sandfish 47, 73, 129
Sand Lizard 50, 55, 154–5
sand microteiids 47

sand racers 153
Saurodactylus 116
Sauromalus 40, 58, 189
saurophagy 66, 68–9
scale organs 30
scales 28–9, 38, 46, 77
scaly-foots 99
Sceloporinae 202–3
Sceloporus 202–3
scent marking 28, 29
Scheltopusik 21, 30, 33, 48, 64, 73, 208
Scincella 132
Scincidae 20–1, 60, 119, 126–39
Scincinae 21, 119, 128–9
Scincomorpha 74, 118–39
Scincus 47, 73, 129
Scleroglossa 12, 31
semaphore geckos 76, 116
sense organs 30–1
seps 20, 123
setae 42–3
sexual dichromatism 48–9
sexual dimorphism 48–9
shade lizards 141, 146–7
Shinisauridae 207, 218
Shinisaurus 218
Shovel-snouted Lizard 37, 38
Side-blotched Lizard 39, 64, 203
Sitana 50, 173
skin 28–9, 70, 74
skinks 7, 20, 21, 22, 31, 43, 47, 52, 54, 56, 60, 70, 73, 74–5, 83, 119, 126–39
skulls 15, 17, 18, 22–5
slender-snouted vinesnakes 23
sliders 21
slow worms 20, 21, 30, 33, 73, 207, 208–9
Smaug 120
Smith's Tropical Night Lizard 60
snake-eyed skinks 119, 136–7
snake-tooth skinks 133
social skinks 119, 130–1
Solomon Islands Monkey-tailed Skink 43
South American legless lizards 21
southern padless geckos 91, 92
Sphaerodactylidae 91, 114–17
Sphaerodactylus 52, 114
Sphenodon 31, 34, 35, 51, 52, 81, 88
Sphenomorphinae 21, 119, 132–3
Sphenomorphus 132
spiny-tailed geckos 74, 95
Spiny-tailed Skink 73
spinytail iguanas 42, 58
Stenocercus 204
sticky-toed geckos 96
Strait of Magellan Lizard 36, 59, 198
Strophurus 74, 95
sun skinks 134–5
supple skinks 119, 138–9
swifts 59, 198, 205
Swollen-snouted Side-blotched Lizard 39, 64, 203

T

tails 6, 19, 43, 73–4, 208
Takydromus 19
Tarentola 33, 111–12
taxonomy 14–17
tegus 30, 141, 144–5
Teiidae 141, 242–5
teiids 60, 141, 142–3
Teiinae 142–3
Teiioidea 141
Tenerife Gallotia 60
Teratoscincus 41, 117
territoriality 50
Tetradactylus 20, 123
Thecadactylus 44–5, 52, 113
thick-tailed geckos 93
Thorny Devil 38, 63, 171
thornytail lizards 204
Tikiguania estesi 8
Tiliqua 31, 56, 60, 70, 130
tongues 30–1, 65
Toxicofera 12
Trachylepis 134
tree skinks 119, 139
Tribolonotus 52, 131
Trioceros 50, 180
Trogonophidae 141, 164–5
Trogonophis 165
Tropiduridae 204
Tuatara 7, 8, 17, 22, 28, 31, 34, 35, 51, 52, 81, 88
Tuberculate Agama 36
Tupinambinae 144–5
Tupinambis 30, 144
turnip-tailed geckos 44–5, 52, 113
Tussock Cool-skink 56
Tympanocryptis 35
Typhlosaurus 47, 127

U

Uma 40
Underwoodisaurus 93
Uranoscodon 41, 204
Uromastycinae 167, 176–7
Uromastyx 40, 59, 73, 176
Uroplatus 70, 106
Urostrophus 196
Uta 39, 64, 203

V

Varanidae 207, 220–31
Varanus 11, 19, 30, 51, 54, 57, 60, 64, 66–7, 80, 220–31
Veiled Chameleons 48
venom glands 6, 12, 73, 207, 217, 230
viviparity 12, 36, 55–6
Viviparous Lizard 36, 55, 155
Voeltzkowia 22
vomeronasal organs 30

W

wall geckos 111
water conservation 38
Water Dragon 170
Weapontail 193
West Indian iguanas 58
whiptails 48, 57, 60, 142–3
wonder geckos 41, 117
woodlizards 193
Woodworthia 59, 96
worm-lizards 7, 8, 17, 18, 20, 25, 35, 46, 52, 64–5, 141, 158–65
writhing skinks 119, 138–9

X

Xantusia 125
Xantusiidae 119, 124–5
Xantusiinae 125
Xenosauridae 207, 213

Z

Zonosaurinae 123
Zonosaurus 123
Zootoca 36, 55, 155
Zygaspis 159

ACKNOWLEDGMENTS

The author would like to thank all the herpetologists, naturalists, and photographers who contributed images for this book, and also Bina Mistry and Steve Slater, who read various drafts and made some valuable comments. Thanks also to Caroline Earle, Wayne Blades, Tom Kitch, and the rest of the Ivy Press team.

PICTURE CREDITS

The publisher would like to thank the following for permission to reproduce copyright material:

l=left; r=right; t=top; b=bottom, c=centre, i=inset.

Alamy Stock Photo: Adrian Davies 33; Albatross 82; Andrew Walmsley 88; Anton Sorokin 199; Auscape/Jean-Paul Ferrero 45l; Biosphoto 164; Biosphoto/Michel Gunther 157; blickwinkel /Hartl 50, /F. Teigler 104, /Trapp 148, /W. Layer 177t; David Davis Photoproductions 166; David Gabis 116; DBI Studio 83; Design Pics Inc/VIBE/Jack Goldfarb 128; Dorling Kindersley Ltd 218, /Frank Greenaway 44; Dpa Picture Alliance Archive 182b; Ernst Mutchnick 230; FLPA 102, 110l, 110r; Frank Hecker 206; imageBROKER/Harry Laub 192, /jspix 144, /Valentin Heimer 30; John Sullivan 216; Ken Griffiths 170; Nature Picture Library 152, /Ben Lascelles 80, /Bence Mate 41, /Chris Mattison 52, 211l, /Daniel Heuclin 229b, /Dave Watts 69, /Laurie Campbell 55, /Michael D. Kern 97, /MYN/Andrew Snyder 150; National Geographic Image Collection/Gabby Salazar 142, /George Grall 158; John Cancalosi 75t; Matthijs Kuijpers 115b; Minden Pictures/Chi'ien Lee 108, 219, /James Christensen 201t, 201b, /Martin Willis 71, /Michael & Patricia Fogden 40t, Pete Oxford 112, Piotr Naskrecki 121; Thomas Marent 123b; Natural History Museum 14r; Papilio/Michael Maconachie 228; Premaphotos 68; Premium Stock Photography GmbH/Frank Teigler 181t; Robbie Labanowski 184; Scott Buckel 203; Stephen Dalton 64–65. **Ardea.com**: Auscape 40b; Biosphoto/Matthijs Kuijpers 181b, /Regis Cavignaux 140; Joseph T and Suzanne L Collins 163; Ken Lucas 161; Science Source/Dante Fenolio 193, /Scott Linstead 172, /Tom McHugh 62b. **Carlos L. de la Rosa** 38. **Chad M. Lane** 212l. **Christian Saavedra/cheloderus** 124. **Conrad Hoskin** 93. **Davide Bonadonna** 9. **E Kustatscher, Naturmuseum Sudtirol** 8. **FLPA**: Biosphoto/Michel Gunther 169, 191b; Imagebroker/Thorsten Negro 182t; Minden Pictures/Tui De Roy 191t. **Frank Glaw** 75B. **Getty Images**: Anders Blomqvist 42; Arto Hakola 200; EyeEm/Floriane Mangiarotti 171; R. Andrew Odum 211r; Zachary Winters 59. **Hans Hillewaert** 208. **Indraneil Das** 89. **Jorge H. Valdez** 39. **Karthik Ak** 178. **Mark O'Shea** 14l, 20, 21, 22, 34L, 35, 43l, 43r, 46, 51b, 56, 57, 117t, 132, 134, 136, 151t, 153, 160, 177b, 189, 202, 210, 215. **Matthew McIntosh** 94. **Matthieu Berroneau** (www.matthieu-berroneau.fr) 129b, 155t, 155b, 156, 223. **Max Jackson** 226. **Mike McCoy** 225b. **Natalie McNear** 125. **Nature Picture Library** (naturepl.com): Barrie Britton 51t; Chris Mattison 131t, 165; Cyril Ruoso 62t; Daniel Heuclin 198, 217; Minden/ Piotr Naskrecki 122; Pete Oxford 183t; Rob Valentic 133, 209; Rod Williams 227; Stephen Dalton 179; Visuals Unlimited 225; Will Burrard-Lucas 31t. **Nik Cole/Mauritian Wildlife Foundation** 81. **Pavel Zuber** 24. **Philippe J. R. Kok** 149. **Ria Winters** 78. **Rob Valentic** 61, 92, 99. **Ryan M Bolton** 54. **Ryan Francis** 95, 98, 100, 229t. **S. Blair Hedges** 162. **Science Photo Library**: Philippe Psaila 114; Stuart Wilson 212r. **Shutterstock**: Andrea Izzotti 105b; Apolla 106t; BLFootage 101; Cathy Keifer 31b; Chantelle Bosch 37 (both); Corina Sturm 45r; Creeping Things 34r; Danny Ye 224; Dave Montreuil 222; David Dohnal 188–189; Ecoprint 103; Eric Isselee 86; Ernie Cooper 76I; dwi putra stock 137; Fabio Maffei 196; fivespots 111, 131b; Federico Crovetto 117b; forest71 135; Frontpage 84; Gudkov Andrey 57, 231; Horst Widlewski 180; Huw Penson 90; Ian Schofield 32b; inkivinki 10; Jan Bures 183b; Jonathan Chancasana 143t; Kristian Bell 47; Kuttelvaserova Stuchelova 6–7; Kurit afshen 118, 139; Luca Nichetti 185t; Marek R. Swadzba 154; Mark Kostich 109b, 115t; Martin Mecnarowski 145t; Matt Cornish 72; Matt Jeppson 73, 143b,; Maridav 190; Martin Pelanek 221; Maximillian Cabinet 36; Milan Zygmunt 168, 188l, 195; Nezvanova 220-221; NickEvansKZN 120; reptiles4all 48, 76m, 77, 96, 109t, 113, 130, 145b, 174, 176, 185b, 187t, 187b, 204, 205, 213, 214; Patrick K. Campbell 146, 147, 151b; Peter Schoeman 123t; Rosa Jay 138, 197; Ryan M. Bolton 107; Sebastian Janicki 29, 32t; Sergey Uryadnikov 66–67; Steve Byland 194; Susan Schmitz 186; Valt Ahyppo 106b; yod67 175. **Steven G Johnson** 11. **Superstock**: Minden Pictures/Stephen Dalton 105t. **Tanmay Haldar** 173. **Thomas Fuhrmann** 49. **Todd W. Pierson** 129t. **Tyrone Ping** 126, 127, 159.

Illustration references:
Tegu skull, 23: Drawn from a photo by Karen E. Peterson in Pough et al, *Herpetology* (2016, 4th ed., Sinauer Associates) (figure 11.16A, p394).
Amphisbaenid skull, 25: Drawn from a photo by the Grant Museum of Zoology in Berkovits & Shellis, *The Teeth of Non-Mammalian Vertebrates* (2017, Academic Press) (figure 6.51, p181).
Squamate skin, 28: Redrawn from Vitt & Caldwell *Herpetology: An Introductory Biology of Amphibians and Reptiles* (2014, 4th ed., Academic Press) (figure 2.14, p52).

All reasonable efforts have been made to trace copyright holders and to obtain their permission for the use of copyright material. The publisher apologizes for any errors or omissions and will gratefully incorporate any corrections in future reprints if notified.